RAND NATIONAL SECURITY RESEARCH DIVISION

Alternative Paths to Korean Unification

Bruce W. Bennett

Prepared for the Korea Foundation

For more information on this publication, visit www.rand.org/t/RR2808

Library of Congress Cataloging-in-Publication Data is available for this publication.
ISBN: 978-1-9774-0183-0

Published by the RAND Corporation, Santa Monica, Calif.

© Copyright 2018 RAND Corporation

RAND® is a registered trademark.

Support RAND

Make a tax-deductible charitable contribution at
www.rand.org/giving/contribute

www.rand.org

Preface

Korean unification is a major issue in both South Korea (officially the Republic of Korea) and North Korea (officially the Democratic People's Republic of Korea). Given the substantial uncertainties about Korea's future, there are a wide variety of paths by which unification might occur, if it occurs. This report is an analysis of potential avenues to unification. The author identifies nine alternatives, analyzes them, and recommends actions that South Korea and the United States can take to follow a more favorable path.

This research was sponsored by the Korea Foundation and conducted within the International Security and Defense Policy Center of the RAND National Security Research Division (NSRD). NSRD conducts research and analysis for the Office of the Secretary of Defense, the Joint Staff, the Unified Combatant Commands, the defense agencies, the Navy, the Marine Corps, the U.S. Coast Guard, the U.S. Intelligence Community, allied foreign governments, and foundations.

For more information on the RAND International Security and Defense Policy Center, see www.rand.org/nsrd/ndri/centers/isdp or contact the director (contact information is provided on the webpage).

Contents

Tables

Summary

There is significant interest in Korean unification in both North Korea (officially the Democratic People's Republic of Korea, or DPRK) and South Korea (officially the Republic of Korea, or ROK). In South Korea, most discussion of unification is based on the assumption that South Korean leaders would control the process because that country's economy and world stature significantly dominate North Korea's. But there are many ways in which unification could occur or be attempted, and each holds vast uncertainties. This report responds to three limitations that I perceive in the traditional literature on Korean unification: (1) a heavy focus on peaceful unification and little discussion of other unification paths, (2) an assumption that unification will succeed once it is started, and (3) a failure to address cases involving unification under North Korean control.

In this report, I outline a series of paths to Korean unification. The paths describe key conditions leading to unification (e.g., conquest of North Korea by South Korea and the United States) rather than detailed scenarios (e.g., how such a war would be won). The paths span three contexts: (1) a major war between North and South Korea, (2) a collapse of the North Korean regime, and (3) a peaceful unification coordinated by North Korea's Kim family regime and the ROK government. In my analysis, I recognize the reality that either North or South Korea might control the unification process (or that control might be shared), that any of these paths might fail to achieve a stable unification outcome, and that there is considerable uncertainty in all of the potential outcomes. To better capture this uncertainty, I examine a series of seven challenges that could dramatically affect both the starting conditions for the paths and the ultimate unifications.[1]

The nine unification paths under the three contexts are as follows: If there is a major war, it could lead to (1) DPRK conquest of South Korea or (2) ROK-U.S. conquest of North Korea.[2] (It could also lead to no unification, but that alternative is excluded from the list because it is not a path to unification.) Also considered is (3) a ROK conquest of North Korea in which South Korea has created improved conditions for unification to reduce the costs of victory and improve postconflict stability. A collapse of the DPRK regime could lead to (4) ROK-U.S. intervention to achieve unification, (5) a negotiated unification with the successor regime, or no unification. A peaceful unification could involve (6) domination by North Korea (North Korea absorbs the South), (7) domination by South Korea (South Korea absorbs the North), (8) full cooperation between the two sides, or (9) full cooperation that creates a confederation dominated by North Korea. The United States would play a supporting role to ROK unification leadership throughout these paths.

Based on the analysis in this report, I conclude that South Korea should avoid many of these paths. Especially in consideration of North Korea's weapons of mass destruction, the paths involving major warfare would likely be disastrous for both South and North Korea. Of the warfare paths, the improved ROK-controlled unification path would be achieved via policies and information operations that mitigate much of the North Korean animosity and would clearly be a preferred option if war happened despite ROK-U.S. preferences. Peaceful North Korean absorption of South Korea would not be favorable for South Korea and is a path that South Korea probably would not agree to, just as peaceful absorption of North Korea is a path that Kim Jong-un would not agree to. Although a peaceful full unification appears favorable at first blush,

---

[1]   Because the paths are intermediate conditions, such as South Korea's conquest of North Korea (with the help of U.S. forces), it is necessary to define the conditions that I expect would exist at the time of the conquest. These would be the starting conditions of a path.

[2]   It is possible that a conflict would result in a partial conquest. Such conquests are considered to be part of the full-conquest paths.

Kim Jong-un would probably find it unacceptable: He either would not agree to it in the first place or would try to isolate North Korea once the initial unification began and then withdraw from it eventually. A DPRK-dominant confederation would likely be more acceptable to Kim Jong-un and give the initial appearance of a successful unification, but this would likely proceed only as far as a loose confederation and then gradually fail. It could also be very costly for South Korea.

This leaves a negotiated unification after North Korea's regime collapses as the likely best path for unification. In fact, several experts on Korea believe that the only path to a stable unification involves regime change in the North followed by a gradual, cooperative, peaceful advance toward unification working with the new DPRK government, likely lasting for many years.[3]

The report concludes with three major recommendations. First, South Korea must avoid using a major war to obtain unification—the cost is no longer acceptable. Second, South Korea needs to develop policies now that would provide most of North Korea's elite with a friendly outcome from unification.[4] Third, South Korea and the United States should counter Kim Jong-un's image in North Korea as a godlike leader and his image in South Korea as a benevolent peacemaker; he is neither. Movement toward unification will likely be impaired if the North Korean and South Korean people believe his propaganda. Still, to avoid an effect opposite that which is desired, this step needs to be done carefully and by experts in psychological operations.

---

[3] See, for example, Bruce E. Bechtol, ed., *Confronting Security Challenges on the Korean Peninsula*, Quantico, Va.: Marine Corps University Press, 2011; David S. Maxwell, "Unification Options and Scenarios: Assisting a Resistance," *International Journal of Korean Unification Studies*, Vol. 24, No. 2, 2015; and Fredrick Vincenzo, *An Information Based Strategy to Reduce North Korea's Increasing Threat: Recommendations for ROK & U.S. Policy Makers*, Washington, D.C.: Center for a New American Security, Georgetown University, National Defense University, and U.S.-Korea Institute at SAIS, October 2016.

[4] Details on how to make unification friendly for North Korean elites can be found in Bruce W. Bennett, *Preparing North Korean Elites for Unification*, Santa Monica, Calif.: RAND Corporation, RR-1985-KOF, 2017.

Acknowledgments

I appreciate the opportunity given to me by the Korea Foundation to examine this extremely important topic. I am grateful for the insights and advice provided by many South Korean colleagues and by several North Korean refugees. These former North Koreans, who had mostly been elites, have provided invaluable insights into their perspectives and the perspectives of their former peers.

I also appreciate the recommendations of Scott Harold and David Maxwell, who served as the reviewers of this report; they have helped me create a much better product.

Abbreviations

DMZ	Korean Demilitarized Zone
DPRK	Democratic People's Republic of Korea
ICBM	intercontinental ballistic missile
KPA	Korean People's Army
kt	kiloton
PLA	People's Liberation Army
psi	pounds per square inch
ROK	Republic of Korea
WMD	weapons of mass destruction

CHAPTER ONE

Introduction

There is not a single, simple path to Korean unification. In 1950, North Korea (officially the Democratic People's Republic of Korea, or DPRK) tried to use military force to achieve unification but failed. The Korean War ended with each side anxious to impose unification on the other but neither able to do so, forcing them into an armistice. Of course, a peaceful unification would be preferable, but the societal and political cultures of the two countries are so different that it is difficult to imagine how a full unification could be peacefully achieved. Even if North Korea or South Korea (officially the Republic of Korea, or ROK) had military dominance, imposing unification could cause so much damage to both sides that the victor would not have the economic resources to support full unification and probably would not have the military forces to stabilize a unified country. South Korea might exploit a collapse of the DPRK regime to achieve unification, but it is difficult to imagine that North Korean elites would accept such a path to unification under current ROK policies.

This report systematically describes several alternative paths to unification and examines their potential for success. It also considers various challenges that would complicate the unification process, making the results of following any given path highly uncertain. In the end, none of the potential paths provides any certainty of a successful unification, and many of the paths could lead to disaster. Moreover, current expectations do not provide the kind of basis required for true success. Both North and South Korea need to adjust their planning to have any significant expectation of success. Most likely, a successful unification will be a long process in which many compromises are made on both sides.

What Does Unification Mean?

There are various forms of Korean unification that could occur. If the DPRK and ROK governments could find a clear common ground, unification could mean a single, integrated government for all of Korea. Alternatively, a weaker form of unification could result in a confederation in which separate DPRK and ROK governments are retained for many issues, but some aspects of the unified Korean government would be shared. The shared aspects might include some combination of foreign policy, national defense, and trade policy.

Unification of Korea could also be geographically full or partial. A geographically partial unification could occur if China intervenes in North Korea and establishes control over some of the Korean territory for a period of time. Unification could also be geographically partial if some parts of Korea refuse to accept the authority of the unified government and successfully resist attempts by the unified government to establish control.

In addition, unification might not be instantaneous. If, for example, ROK and U.S. forces go into North Korea to impose unification after a collapse of the DPRK government, unification could be a rolling process in which some DPRK areas along the Korean Demilitarized Zone (DMZ) are brought into the unified Korea immediately while other areas in North Korea are added to the unified Korea as they are freed or captured. And even when these areas are brought into the unified Korea, the DPRK people—and especially the DPRK military and security services—might not accept the authority of the unified Korean government and might rebel against it. Therefore, contrary to commonly held assumptions, stabilization must begin simultaneously with the military advance of ROK and U.S. forces into North Korea.

Thus, I refer to achieving initial unification in terms of possession of territory. Thereafter, the key issue is whether the unified Korean government is able to establish stabilized control over the territory it possesses. There are various scenarios in which the ROK government could establish initial control over significant parts of North Korea but not be able to stabilize those areas. This lack of stability could become so serious that it would cause the unification to fail.

Indeed, if a serious and costly insurgency were to develop in the occupied territories, the ROK government could decide to abandon its efforts at unification.[1]

The Purpose of This Report

This report examines the prospects for these alternative paths. There is already a very rich literature on Korean unification,[2] but in my view, that literature could be improved upon by examining a larger number of potential unification paths and their plausibility. In particular, the literature is often limited in three ways:

1. It tends to focus on peaceful unification despite the other possible contexts in which unification could occur.
2. It usually assumes that unification will proceed successfully once it begins.
3. It almost always ignores the potential for a DPRK-led unification.[3]

---

[1]   In these examples, I have used the ROK government as the organization establishing the unified Korea. But as will be discussed shortly, I also include unification paths in which the DPRK government establishes control over the unified Korea. There is relatively little talk of such cases in South Korea and the United States, but when Kim Jong-un speaks of unification, it is often clear that he is speaking of a unification controlled by North Korea.

[2]   Because I do not speak Korean, I am familiar with the English literature only, but even that is voluminous. See, for example, David F. Helvey, *Korean Unification and the Future of the U.S.-ROK Alliance*, Washington, D.C.: National Defense University, Strategic Forum 291, February 2016; Robert Kelly, "The German-Korean Unification Parallel," *Korean Journal of Defense Analysis*, Vol. 23, No. 4, December 2011; Andrei Lankov, "A Legal Minefield for Korean Reunification," *Asia Times*, July 30, 2011; David S. Maxwell, "Unification Options and Scenarios: Assisting a Resistance," *International Journal of Korean Unification Studies*, Vol. 24, No. 2, 2015; Park Geun-hye, "A New Kind of Korea," *Foreign Affairs*, September/October 2011; Jonathan D. Pollack and Chung Min Lee, *Preparing for Korean Unification: Scenarios and Implications*, Santa Monica, Calif.: RAND Corporation, MR-1040-A, 1999; Jochen Prantl and Hyun-wook Kim, "Germany's Lessons for Korea: The Strategic Diplomacy of Unification" *Global Asia*, December 27, 2016; ROK Ministry of Unification, *Initiative for Korean Unification*, Seoul, October 2015; and ROK Ministry of Unification, *2016 White Paper on Korean Unification*, Seoul, May 2016.

[3]   For example, this is true for all of the works cited in the previous footnote.

This report describes nine alternative paths to unification that seek to go beyond these limitations. My first concern is with the literature's focus on peaceful unification. That focus is seen especially in ROK government literature that addresses unification but also appears in the academic literature.[4] For example, in 2002, the ROK Ministry of Unification wrote about the Sunshine Policy for Peace and Cooperation,[5] consistent with President Kim Dae-jung's Sunshine Policy and summit meeting with North Korean leader Kim Jong-il, which won Kim Dae-jung the Nobel Peace Prize.[6] Former ROK President Park Geun-hye made a major issue of adjusting South-North relations toward achieving unification. She wrote a related article for *Foreign Affairs* before she was elected president.[7] After her inauguration, she spoke (during her New Year's press conference on January 6, 2014) of unification being an economic bonanza for South Korea and made a major speech on unification policy in Dresden, Germany, on March 28.[8] Under President Park, the ROK Ministry of Unification provided clear proposals on how a peaceful unification should proceed.[9] In contrast, President Moon Jae-in has had his ROK Ministry of Unification take a different approach, with a focus on peaceful coexistence with North Korea and

---

[4]   See, for example, Byeonggu Lee, "The Role of the Republic of Korea-U.S. Alliance in Peaceful Unification of Korea," *International Journal of Korean Studies*, Vol. 19, No 2, Fall 2015.

[5]   ROK Ministry of Unification, *Sunshine Policy for Peace and Cooperation*, Seoul, May 2002.

[6]   Then-President Kim Dae-jung started his presidency in 1998 by stating in his inaugural address, "First, we will never tolerate armed provocation of any kind. Second, we do not have any intention to harm or absorb North Korea. Third, we will actively push reconciliation and cooperation between the South and North beginning with those areas which can be most easily agreed upon." This approach was clearly peaceful and was intended to nudge North Korea toward reconciliation and eventual unification. See Kim Dae-jung, "Inaugural Address," Ch'ongwadae, South Korea, February 25, 1998.

[7]   Park, 2011.

[8]   Kim Myong-sik, "'Unification Bonanza' Is Misleading Slogan," *Korea Herald*, February 5, 2014; "Dresden Initiative for Peaceful Unification on the Korean Peninsula," KBS World Radio, March 28, 2014.

[9]   ROK Ministry of Unification, 2015; ROK Ministry of Unification, 2016.

the vague and far future goal of "peaceful unification through mutually beneficial cooperation."[10]

But there are other outlooks. RAND Corporation colleagues Jonathan Pollack and Chung-min Lee wrote in 1999 about four potential unification contexts consisting of the three used herein (war, collapse, and peace) and a fourth that is a variant of collapse in which China intervenes significantly.[11] Evans Revere argues that, "Despite the U.S. and ROK preference, peaceful reunification is by no means guaranteed. Reunification reached as a result of North Korean collapse, internal chaos in the North, or civil war would impose a massive burden on the Korean people as they seek to build a united nation."[12] The *New York Times* Editorial Board argues,

> In the best case, peaceful reunification would reunite long-separated families, free 24 million North Koreans from dictatorship, enhance regional security and eliminate North Korea's nuclear threat. Unfortunately, other outcomes seem more likely: the continuation of the present hostile impasse, or, conceivably, the violent collapse of the North Korean regime.[13]

My second concern is with the usual assumption that unification will proceed successfully once begun. This assumption underlies both former President Park's claim that unification would be an economic bonanza and efforts to evaluate that bonanza.[14] Many studies have

---

[10] ROK Ministry of Unification, "Moon Jae-In's Policy on the Korean Peninsula: A Peninsula of Peace and Prosperity," Seoul, 2017.

[11] This report considers Chinese intervention as possible in any of the three contexts considered and identifies it as a key challenge. See Pollack and Lee, 1999. A subsequent study (Maxwell, 2015) considered four contexts, splitting the regime collapse context herein into internal regime change and regime collapse.

[12] Evans J. R. Revere, "Korean Reunification and U.S. Interests: Preparing for One Korea," Washington, D.C.: Brookings Institution, January 20, 2015.

[13] "Is Peaceful Korean Unification Possible?" editorial, *New York Times*, December 11, 2014.

[14] See, for example, "What North and South Korea Would Gain If They Were Reunified," *The Economist*, May 5, 2016.

assumed that a unification would be a success and examined conditions after unification.[15]

Again, there are different views. Austin Long argues,

> The potential for an insurgency beginning after the collapse of the DPRK appears contingent but significant. The ROK should plan accordingly, particularly in terms of how to treat elements of the former KPA [Korean People's Army] and other mid- and low-level government officials of the DPRK. Substantial numbers of troops will be needed for an extended period of time just to secure weapons and demobilize the KPA even in the best circumstances.[16]

My own work talks about the multiple potential unification challenges after a potential North Korean government collapse,[17] as well as the implications of not including the interests of North Korean elites in unification.[18] But even in these works, there is often not an examination of a failed unification: What might lead to South Korea aborting a unification and leaving North Korea in chaos?

My final concern is with the lack of discussion of DPRK-led unification. I have discussed the possibility of a DPRK-led unification with ROK colleagues many times, and they adamantly dismiss it as impossible. The basic argument is that, "Given the South's economic prowess, alliance with the United States, and international influence . . . South Korea would likely absorb and integrate the North, not the other way

---

[15] For example, an examination of the U.S.-ROK alliance after a unification is found in Helvey, 2016.

[16] Austin Long, *Insurgency in the DPRK? Post-Regime Insurgency in Comparative Perspective,* Baltimore, Md.: U.S.-Korea Institute Paul H. Nitze School of Advanced International Studies, Johns Hopkins University, March 2017.

[17] Bruce W. Bennett, *Preparing for the Possibility of a North Korean Collapse,* Santa Monica, Calif.: RAND Corporation, RR-331-SRF, 2013.

[18] Bruce W. Bennett, *Preparing North Korean Elites for Unification,* Santa Monica, Calif.: RAND Corporation, RR-1985-KOF, 2017.

around."[19] Some previous work simply assumes that a unification will be ROK-led without such an explicit statement.[20]

But DPRK-led unification was clearly Kim Jong-un's interest in his 2018 New Year's Day address, when he spoke of unification a dozen times and proposed hastening unification, emphasizing that unification is a key goal of all Koreans.[21] And some literature does consider DPRK-led unification.[22] Admittedly, this would be an unhappy outcome for South Korea and the United States. But it would also likely be an unhappy outcome for Kim Jong-un and his regime—a key point that Kim's enthusiasm for leading unification seems to miss.

Addressing these three concerns simultaneously leads to a very different perspective on South Korean and U.S. policy toward Korean unification. This perspective recognizes the need to prepare now for a successful unification because current preparations are inadequate. In recent years, there has been some significant literature to address these concerns and help inform the work presented in this report.[23]

---

[19] Sue Mi Terry, *Unified Korea and the Future of the U.S.-South Korea Alliance*, New York: Council on Foreign Relations, December 2015, p. 3.

[20] This is true of the author's work on potential North Korean collapse, as well as the excellent work of Victor Cha and David Kang, *Challenges for Korean Unification Planning: Justice, Markets, Health, Refugees, and Civil-Military Transitions*, Washington, D.C.: USC Korean Studies Institute and Center for Strategic and International Studies, December 2011.

[21] Kim Jong-un, "New Year's Address," NK Leadership Watch, January 1, 2018.

[22] Lee, Sung-yoon, "North Korea's Revolutionary Unification Policy," *International Journal of Korean Studies*, Vol. 18, No. 2, Fall 2014.

[23] See, for example, Bruce E. Bechtol, Jr., ed., *Confronting Security Challenges on the Korean Peninsula*, Quantico, Va.: Marine Corps University Press, 2011; David S. Maxwell, "A Strategy for the Korean Peninsula: Beyond the Nuclear Crisis," *Military Review*, September/October 2004; David S. Maxwell, "Should the United States Support Korean Unification and If So, How?" *International Journal of Korean Studies*, Vol. 18, No. 1, Spring 2014; and Fredrick Vincenzo, *An Information Based Strategy to Reduce North Korea's Increasing Threat: Recommendations for ROK & U.S. Policy Makers*, Washington, D.C.: Center for a New American Security, Georgetown University, National Defense University, and U.S.-Korea Institute at SAIS, October 2016.

Methodology

This project draws on my knowledge and expertise, as well as interviews and discussions with ROK government leaders and DPRK refugees, ROK and U.S. experts, and the news media. The unification paths recommended stem from contexts discussed by others, but I have refined them based on extensive thinking about Korean unification and my review of these concepts with experts on Korea. The analysis of the paths is based on my understanding of ROK and DPRK attitudes toward unification and the various conditions that would be associated with unification, including government control, information flow, opposition to imposed conditions, and so forth. These conditions also were discussed with ROK, DPRK, and U.S. experts to refine the results. In addition, I have participated in quantitative modeling and wargaming that has examined some of the unification paths and some of the contexts, such as a DPRK invasion of South Korea, and I use the results of these experiences as part of the basis for analyzing the paths.

The framework for analysis in this report involves three components. The first is a description of the contexts in which unification could occur and the unification paths associated with these contexts. The second component is a description of the challenges that could occur in the unification process. The third component is an analysis of the likely outcomes of each path, including the feasibility and probability of the paths.

Unification Contexts and Associated Paths

I have talked and written much about unification contexts for years,[24] generally describing three contexts:

1. a major war between North and South Korea
2. a collapse of the North Korean regime
3. a peaceful unification coordinated by the Kim family regime and the government of South Korea.

---

[24] See, for example, Bennett, 2013; Bennett, 2017.

However, I have not previously tried to characterize the alternatives more systematically; doing so is the subject of this report.

Most open discussion of Korean unification occurs in South Korea and focuses on ROK-controlled paths to unification. But, as noted already, that is not what the leaders of North Korea—Kim Jong-un and his father and grandfather before him—have talked about. They have focused on DPRK-controlled unification, and especially on the conquest of South Korea. A North Korean invasion of South Korea would be a follow-up to its failure to achieve this objective in the Korean War of 1950 to 1953. A DPRK-controlled unification would potentially be possible in either the first or third of the listed contexts.

With regard to the warfare context, a limited war is unlikely to lead to unification and is thus excluded from consideration.[25] The same is true for major wars in which neither side dominates, because neither side would be able to accomplish unification. Therefore, the two paths presented begin from conditions in which either North Korea dominates the war outcomes or South Korea dominates. From those paths, I then examine how likely it is that the intermediate condition would develop (e.g., that the North would conquer the South or the South, with support from the United States, would conquer the North), how it might lead to unification, and the likelihood that it leads to unification.

Control of the unification process is a major factor used in identifying different paths. For example, in the peaceful unification context, there is the possibility that North and South Korea could achieve a unification with balanced control. A balanced unification would require significant adjustments to the government style and overall culture for at least one of the two countries, and possibly both. Is it likely that the leaderships of both countries would agree to a balanced unification? If so, how might it develop? And what is the potential for that unification to develop and stabilize?

---

[25] Many of my ROK military colleagues feel that any artillery fired at Seoul would not be a limited conflict but rather the beginning of a major war. Of course, that could change as the situation develops, but, based on gaming of limited conflicts that I directed at RAND in December 2016, the potential for a rapid escalation spiral is fairly high in Korea.

The combination of the contexts and the control of the unification process define the paths to unification and some alternatives. These are developed in Chapters Three (on war), Four (on collapse), and Five (on peaceful unification).

Unification Challenges

There are many potential challenges to Korean unification. These challenges would affect and potentially dominate any of the paths to unification and the outcomes of those paths. For the purposes of this report, seven key challenges are considered:

1. North Korean weapons of mass destruction (WMD)
2. other North Korean military forces
3. reduced size and readiness of the active-duty ROK Army
4. inadequate training of ROK Army reserve forces
5. slow deployment of U.S. forces to Korea
6. inadequate ROK planning for a replacement government in North Korea
7. Chinese intervention in North Korea.[26]

These challenges are described in Chapter Two and then applied to the unification paths in Chapters Three, Four, and Five.

There are, of course, other challenges that could be considered in assessing the paths to unification described herein. For example, the age of most North Korean military equipment and North Korea's apparent lack of adequate military supplies could affect the unification paths. ROK politics and the divisions between the conservatives and the progressives could also affect the paths. After considering other challenges, I determined that the seven aforementioned challenges are the most significant.

Paths to Unification

Most experts considering Korean unification describe one to three paths that are fairly set and certain. Even those considering the uncer-

---

[26] These challenges are not sequenced by the level of expected impact.

tainties in Korean unification often define a baseline path with little consideration of the seven challenges described in the previous section. This report examines nine unification paths under three contexts. In each context, the potential feasible paths are considered. Thus, war could lead to North Korean conquest of South Korea, South Korean and U.S. conquest of North Korea,[27] or no unification. (The third alternative is not a path to unification and is thus excluded.) Also considered is a ROK conquest of the North in which South Korea has created improved conditions for unification to reduce the costs of victory and improve postconflict stability. A collapse of the North Korean regime could lead to ROK-U.S. intervention to achieve unification, a negotiated unification with the successor regime, or no unification. A peaceful unification could be represented as domination by North Korea (North Korea absorbs the South), domination by South Korea (South Korea absorbs the North), full cooperation between the two sides, or full cooperation that creates a confederation dominated by North Korea. By context, these paths and the labels used in this report to describe each path (in parentheses) are as follows:

- a major war between North Korea and South Korea
    1. unification via North Korean conquest of the South (*War: DPRK conquest*)
    2. unification via South Korean and U.S. conquest of the North based on the current ROK-U.S. approach, which would yield a costly victory and a contested peace (*War: ROK-U.S. conquest—costly victory, contested peace*)
    3. improved ROK-controlled unification resulting from major war, intended to reduce the costs of victory and improve the outcomes (*War: Improved ROK-U.S. conquest—reduced costs of victory, improved postconflict stability*)
- a collapse of the North Korean regime
    4. ROK-U.S. military intervention in North Korea (*DPRK collapse: Intervention*)

---

[27] It is possible that a conflict could result in a partial conquest. Such conquests are considered to be part of the full-conquest paths.

   5. unification negotiated with Kim Jong-un's successor(s) (*DPRK collapse: Negotiation*)

- a peaceful unification coordinated by the Kim family regime (with Kim Jong-un still in control of North Korea) and the ROK government
   6. North Korea absorbs the South (*Peace: DPRK absorption*)
   7. South Korea absorbs the North (*Peace: ROK absorption*)
   8. full cooperation between North and South (*Peace: Cooperation*)
   9. North Korea–dominant confederation (*Peace: North Korea dominates*).

The United States would play a supporting role to ROK unification leadership throughout these paths.

Organization of This Report

This report consists of six chapters. As noted earlier, Chapter Two addresses the challenges that would probably complicate Korean unification. Chapter Three describes and analyzes the paths to Korean unification that could be associated with a major war between North and South Korea. Chapter Four describes and analyzes the paths to Korean unification that could be associated with a North Korean regime collapse. Chapter Five describes and analyzes the paths to peaceful Korean unification of the Kim family regime and the government of South Korea. And Chapter Six summarizes the assessments of the previous chapters and provides recommendations. The appendix provides more details on the challenges associated with North Korean WMD.

CHAPTER TWO

Potential Unification Challenges

Before turning to the unification paths, I begin by discussing the challenges that would substantially affect the paths. Most ROK and U.S. discussions of Korean unification assume that unification will be achieved peacefully or that, if force is used, it will not complicate achieving a peaceful outcome. In practice, the societal and cultural differences between North Korea and South Korea make conflict a significant possibility en route to an initial unification, and these differences might require substantial use of force to achieve stability and an acceptable, long-term full unification. These confrontations could significantly challenge a full unification, as could other key challenges.

This chapter describes the challenges that could impair successful unification across the nine alternative paths. I begin by examining the seven major challenges that have the potential to impair unification. As outlined in Chapter One, these challenges are

1. North Korean WMD
2. other North Korean military forces
3. reduced size and readiness of the active-duty ROK Army
4. inadequate training of ROK Army reserve forces
5. slow deployment of U.S. forces to Korea
6. inadequate ROK planning for a replacement government in North Korea
7. Chinese intervention in North Korea.

I then turn to examining why even a theoretically peaceful unification likely will still require some use of military force.

North Korean Weapons of Mass Destruction

North Korea deploys a combination of nuclear, chemical, and biological weapons and might also plan to employ radiological weapons; all of these pose severe threats. The appendix to this report looks in more detail at these threats. This section discusses how the effects of WMD could affect unification.

Nuclear Weapons

Estimates range widely on how many nuclear weapons North Korea possesses. A Council on Foreign Relations study in 2017 summarized this uncertainty: "Estimates of the country's nuclear stockpile vary: some experts believe Pyongyang has between fifteen and twenty nuclear weapons, while U.S. intelligence believes the number to be between thirty and sixty bombs."[1] Many of these would still likely be of the smaller 10-kiloton (kt) destructive power that North Korea tested through 2016, although the portion of larger-yield weapons (like the 250-kt weapon tested in 2017) will almost certainly increase over time. One media report talks of "a new U.S. Defense Intelligence Agency (DIA) assessment that North Korea may be accruing fissile material at a rate sufficient to add 12 nuclear weapons to its growing arsenal each year."[2] If the growth rate is somewhere between six and 12 weapons per year, then the estimate of 15 to 60 in 2017 transforms to roughly 30 to 100 in 2020. This growth in the threat is part of what appears to be motivating U.S. and ROK efforts to dismantle the North Korean nuclear weapon program.

Would North Korea use these weapons against ROK cities? A historical event sheds light on this question. During the first nuclear crisis with North Korea in 1993, Kim Jong-un's grandfather, Kim Il-sung, assembled his senior military personnel to talk about the possibility of having to fight the United States over the North's nuclear weapon pro-

---

[1]   Eleanor Albert, "North Korea's Military Capabilities," Council on Foreign Relations, September 5, 2017.

[2]   Ankit Panda, "US Intelligence: North Korea May Already Be Annually Accruing Enough Fissile Material for 12 Nuclear Weapons," *The Diplomat*, August 9, 2017.

gram. He undoubtedly surprised the military personnel by asking them what they would do if the North lost such a conflict. Kim Jong-un's father responded: "I will be sure to destroy the Earth! What good is this Earth without North Korea?"[3] Kim Il-sung responded by praising his son as having identified the correct response—a response that appears to still be planned today.[4] North Korean nuclear weapons like the first five it tested could cause hundreds of thousands of deaths and serious casualties; a weapon like its sixth test could cause millions of deaths and serious casualties.[5] Attacks on cities would certainly destroy much of the economy and infrastructure of South Korea, potentially leaving it unable to perform any kind of Korean unification.

DPRK nuclear weapons also jeopardize ROK and U.S. military forces. According to the Korea Institute for Defense Analyses, "North Korea threatened that Pyeongtaek is 'our military's foremost strike target.' Such strong response from North Korea indicates that North Korea perceives Camp Humphreys as a grave threat."[6] A 250-kt North Korean nuclear weapon could destroy most of Camp Humphreys (or Osan Air Base, both of which are U.S. military bases in South Korea), largely eliminating or incapacitating the people, equipment, and supplies located there. Some North Korean defectors argue that Kim Jong-un would target U.S. military facilities in South Korea hoping to kill 20,000 or so Americans, with the expectation that the United States could not accept such losses and would withdraw from the conflict. The appendix provides more details.

---

[3] Kim Hyun Sik, "The Secret History of Kim Jong Il," *Foreign Policy*, September/October 2008.

[4] According to several North Korean defectors with whom I have spoken, this saying is still posted in conspicuous places in Pyongyang and thus likely expresses one component of how North Korea might use its nuclear weapons.

[5] See the appendix for more details of the damage that could be caused.

[6] Park Won Gon, "Strategic Implications of the USFK Relocation to Pyeongtaek," *ROKAngle*, No. 164, Korea Institute for Defense Analyses, October 20, 2017, p. 4.

Other Weapons

North Korea possesses substantial quantities of chemical weapons and unknown quantities of biological and radiological weapons.[7] These weapons could cause substantial damage in South Korea, as discussed in more detail in the appendix.

Threat to Korean Unification

North Korean WMD could negatively affect Korean unification in various ways. WMD could cause so much damage to South Korea, North Korea, or both that the two countries would be unable to achieve a unification. And even if they did, the unified Koreas would lack the resources necessary to stabilize and begin rebuilding, ultimately leading to a less successful unification that could cause a unified Korea to collapse. Massive external assistance would be required and would need to start with military intervention in the North to limit the humanitarian disaster that would otherwise develop. In the event of North Korean WMD use, South Korea would almost certainly lack the resources to provide the needed humanitarian aid even to South Koreans, let alone to North Koreans—potentially causing anger among South Koreans toward North Koreans. North Korean WMD use would also seriously alienate people in South Korea, pushing them to potentially demand reparations from the North Korean people. South Koreans might also be induced to seek revenge against North Korea and its people, and North Koreans might reciprocate. The resulting conflict could cause Korean unification to fail.

Other North Korean Military Forces

North Korea fields a very large non-WMD military force, as illustrated in Table 2.1. In a war, this large force will make North Korean advances more feasible or allow North Korea to better defend itself. Thus, these DPRK military forces should significantly affect whether

---

7   On chemical weapons, see ROK Ministry of National Defense, *2016 Defense White Paper*, December 31, 2016, p. 34.

Table 2.1
South and North Korean Conventional Military Forces, 2016

Type of Forces	South Korea	North Korea
Army active-duty personnel	490,000	1,100,000
Total active-duty personnel	625,000	1,270,000
Reserve personnel	3,100,000	7,620,000
Tanks	2,400	4,300
Armored vehicles	2,700	2,500
Artillery, multiple rocket launchers	5,900	14,100
Combat aircraft	410	810
Surface combatants	110	430
Submarines	10	70

SOURCE: ROK Ministry of National Defense, 2016.

the North conquers the South or the South conquers the North—and thus the path in the war context.

The DPRK conventional forces would also affect the ability to achieve a stable unification. In several of the unification paths identified in this report, South Korea and the United States must find a means to demilitarize a very large number of North Korean personnel who would no longer be needed in the military after Korean unification.[8] Stabilization of unifications that are not led by North Korea will require handling these personnel. If angered by the circumstances of warfare or another path to unification, these DPRK military personnel could rebel against the unified Korean government. Thus, to pacify these personnel, the unified government must at least find jobs

---

[8] Even in the warfare or North Korean collapse contexts, the ROK and U.S. objective will not be annihilation of all North Korean military personnel but rather destruction of the North Korean military as a coherent body. The North Korean personnel will retain military skills and likely know where to find weapons throughout North Korea, which would pose an insurgent problem if it is not pacified.

for them to minimize their potential for rebellion.[9] This will be particularly true with the reportedly 200,000 North Korean special forces and likely 200,000 or so security services personnel who would be well trained in disruptive attacks.[10]

Reduced Size and Readiness of the Active-Duty ROK Army

As shown in Table 2.1, the ROK Army in 2016 amounted to 490,000 active-duty personnel; the remaining 135,000 ROK military personnel fill billets in the ROK Air Force and ROK Navy (including the ROK Marine Corps). ROK President Moon has committed to reducing the ROK active-duty military to 500,000 personnel by the time that he leaves office in 2022; the reductions will come partly because of demographics and partly because President Moon wants to reduce the length of the conscription period.[11] In recent years, the ROK Army has already been reduced from 560,000 active-duty personnel, while no changes have been made to the personnel in the ROK Air Force or ROK Navy (except for a modest increase in the size of the ROK Marine Corps after the shelling of Yeonpyeong Island in 2010).[12] If the ROK Air Force and the ROK Navy remain with a total of 135,000 active-duty personnel, then the ROK Army might shrink to just 365,000 active-duty personnel in 2022.[13] South Korean

---

[9] See Long, 2017, p. 17; Bennett, 2013.

[10] On special forces, see ROK Ministry of National Defense, 2016, p. 29.

[11] According to a *Chosunilbo* article, "The Defense Ministry pledged Friday to slash the number of troops from the current 620,000 to 500,000 and shorten mandatory military service from the current 21 months to 18 by 2022" (Jun Hyun-suk, "Troops to Be Slashed to 500,000 by 2022," *Chosunilbo*, January 22, 2018).

[12] This is a comparison of information in the 2016 and 2000 ROK Defense White Papers. See ROK Ministry of National Defense, 2016, p. 268; and ROK Ministry of National Defense, *2000 Defense White Paper*, Seoul, 2000, p. 272.

[13] Song Sang-ho, "The Cut Will Come in Sync with Plans to Pare Down the Number of Active-Duty Troops to 500,000 by 2022 from the Current 618,000," Yonhap News Agency, July 27, 2018. The 500,000 personnel would be for all services. The Defense Reform Plan

demographic trends could push this number down further to roughly 300,000 by 2027.[14]

As previous work has demonstrated, a ROK Army of 365,000 personnel likely would not be sufficient to stabilize North Korea after a relatively peaceful unification, even with some U.S. assistance.[15] In the unification paths in the war context, a ROK Army of this size is unlikely to be sufficient to defeat the North Korean military and create the conditions for ROK-led unification, let alone accomplish stabilization. Thus, even if Korean unification could somehow be achieved, such a unification could well fail because of North Korean insurgency that the ROK military could be too small to control.

One option for dealing with manpower shortages would be to co-opt some of North Korea's KPA forces. These forces could be used under ROK supervision and could help reassure the North Korean population. But this approach also poses risks in areas where North Korean personnel have a history of stealing food or otherwise misbehaving. ROK and U.S. personnel would need to closely monitor the KPA units and impose strong discipline on them.

As part of the Singapore Summit meeting between U.S. President Donald Trump and Kim Jong-un in June 2018, Trump agreed to

---

reduces the manpower only in the ROK Army. Thus, the ROK Air Force and Navy will remain at 135,000 personnel, making the ROK Army 365,000. See ROK Ministry of National Defense, 2016, p. 268.

[14] This result is estimated from my Excel spreadsheet model that calculates the size of the ROK military based on the size of the age cohort available to volunteer or be drafted and various other factors. The ROK age cohort for the draft is 20-year-old men, and that cohort will fall from 271,000 in 2022 to 227,000 in 2026 before rebounding and then falling again. See Korean Statistical Information Service, "Projected Population by Age (Korea)," population database, undated.

[15] In an attempt to place a lower bound on the stabilization forces required, colleague Jennifer Lind and I described a condition in which, after a DPRK government collapse, the numerically larger faction requests ROK absorption of the North. Even then, "260,000–400,000 ground force personnel would be required to stabilize North Korea." See Bruce W. Bennett and Jennifer Lind, "The Collapse of North Korea: Military Missions and Requirements," *International Security*, Vol. 36, No. 2, Fall 2011. The more realistic paths considered in this report would not have stabilization conditions as good as postulated in that paper. Moreover, any Army has a substantial overhead of administrative personnel and training personnel who would not be available for stabilization.

suspend major ROK-U.S. training exercises. This suspension has been something the North Koreans have demanded for decades. But these exercises have played a major role in training the ROK and U.S. forces on how to defend against a North Korean invasion of South Korea. As the experience of these exercises is lost, the readiness of ROK and U.S. forces will degrade, especially because these exercises were used to build a combined ROK and U.S. military team and develop an understanding for how the ally would operate.

Inadequate Training of ROK Army Reserve Forces

The low number of anticipated ROK Army active-duty troops means that the ROK Army needs a more capable reserve force, especially for offensive and stabilization operations. Although the ROK Army has a large reserve force, as shown in Table 2.1, those personnel receive a maximum of only three days of training per year.[16] Historically, three days might have been sufficient for soldiers who would need to stand north of Seoul and fire at invading North Korean personnel, especially because the vast majority of those ROK personnel had served in active-duty military units. But offensive operations and stabilization require a considerably greater level and range of military skills and substantial military unit cohesion, and three days of training per year is a small fraction of what would be required to prepare those personnel.

The ROK military has recognized this problem, but political constraints have prevented a solution. As of 2010, the ROK military goal was to increase the annual training to a maximum of four or five days, still only a small fraction of what is needed. I have proposed that the ROK Army needs two options for reserve training: (1) training of defensive forces in the current manner but with additional training each year and (2) more-substantial training, similar to that of the U.S. Army reserves, of one weekend a month and two weeks each summer for service in reserve divisions that would specifically have stabilization

---

[16] ROK Ministry of National Defense, 2016, p. 226.

roles.[17] As ROK youths enter the military reserve force, they should be given a choice between these two tracks. The second track would become the manpower pool from which reserve combat units would be constituted to support offensive operations into North Korea and the stabilization of that area. Those choosing the second track should be awarded a scholarship to cover their college tuition until they graduate, including studies at the graduate level.[18] The second track would increase Korean academic degrees by making studies more feasible and thereby raise the education level of the Korean populace; it would be a social program with a military training requirement (because the ROK military could not afford the tuition costs).

The performance of the reserve force can also be improved by training all active-duty personnel (including ROK Air Force and Navy personnel) for stabilization operations while they serve on active duty, making them more prepared to someday perform stabilization operations. Then, when reservists report to their units, they will already have the basic stabilization training needed.

Slow Deployment of U.S. Forces to Korea

For decades, the U.S. military has understood that it needs to deploy promptly to Korea to provide adequate assistance to the ROK military in any major combat situation. To facilitate such prompt deployments, the United States has pre-positioned equipment for some units in Korea; established key command, control, communications, and intelligence capabilities in Korea; and established some of the needed logistical support in Korea for follow-on units. For example, the U.S. Army maintains almost 20,000 personnel in South Korea, although that number includes only one U.S. combat brigade of 4,500 or so personnel.

---

[17] Bennett, 2013, pp. 290–294. The ROK Defense Reform Plan has prepared to increase annual reserve training to four days and eventually five days, which is more than three days but not nearly enough to establish the unit cohesion needed for stabilization.

[18] The scholarship options for ROK Army reservists is something I have been discussing with ROK Army leadership for about four years.

Even with these preparations, deploying U.S. forces to Korea takes considerable time. U.S. Air Force combat units can usually deploy most quickly, with estimates showing that many of those units could fly to Korea within days as long as the airfields planned to support them have not been subjected to attacks using WMD. Many U.S. Navy combat ships and U.S. Army and Marine Corps combat units would take weeks or months to deploy to Korea, assuming undamaged reception ports and airfields. More time would be required to fully deploy the logistics associated with these combat forces. Thus, ROK military units must be prepared to bear the vast majority of defensive combat responsibilities during the initial weeks and months of major combat in Korea or during the early period of ROK and U.S. intervention in North Korea to deal with the North Korean government collapse or other disruptive circumstances.

This pace of U.S. military deployments could prove disruptive for Korean unification in various ways. The combination of North Korean WMD use and large conventional DPRK forces could overwhelm ROK defenses, especially once the ROK Army falls to 365,000 active-duty personnel—or, even worse, to 300,000. Even if the ROK and U.S. defenses remain cohesive, ROK Army units could suffer unacceptable attrition, leaving them unable to operate effectively in subsequent phases of combat operations to achieve unification. North Korea might even conclude that the use of WMD for preemptive destruction of the major U.S. bases in Korea could convince the United States to abandon its role in Korea. The United States needs to be very clear that that is not the case and that any such North Korean action would likely lead to a major U.S. escalation, causing massive destruction to the DPRK regime, military, and security services.

Inadequate ROK Planning for a Replacement Government in North Korea

For decades, the DPRK regime has indoctrinated its people and especially its elites to believe that the United States is their eternal enemy and that South Korea is a U.S. lackey. The regime has also made the

United States and South Korea its scapegoats for its many failings. And to avoid any rebellion by the elites, the regime has sought to convince them that U.S. and ROK hostility would doom them if the regime lost control of the North. The fate of the elites in East Germany after German unification has helped strengthen this point.[19]

There has also been a failure by South Korea to plan for a good future for North Korean elites. Instead, there have been ROK discussions of criminal action against these elites after unification. In addition, South Korea has chosen ROK citizens to become the governors, mayors, and other government officials in North Korea upon unification, clearly planning to remove the North Korean officials.[20] Many North Korean officials can thus be expected to resist unification under ROK leadership. Indeed, how could any kind of peaceful unification occur if those with power in the North would be the victims of such a unification?

Chinese Intervention in North Korea

The Chinese leadership has been fairly clear that, in various circumstances, it might intervene in Korea.[21] For example, Chinese leader Xi Jinping has said, "As a close neighbor of the peninsula, we will absolutely not permit war or chaos on the peninsula. This situation would not benefit anyone."[22] As a leader who cannot prevent North Korean missile tests, Xi might seek to deal with war or chaos on the peninsula by inserting Chinese military forces. One author speaks of this concept as the *Xi doctrine*: "This 'doctrine' appears to reserve to China the right

---

[19] These points are made in some detail in Bennett, 2017, pp. 13–14.

[20] These ROK individuals are part of the Commission of the Five North Provinces, as discussed in Bennett, 2017.

[21] I do not speak Chinese and thus have only very limited access to Chinese texts. The observations in this section are largely limited to Chinese perspectives that I have read in the English-language literature.

[22] Michael Martina, "China Won't Allow Chaos or War on Korean Peninsula: Xi," Reuters, April 28, 2016.

to use force to intervene in conflicts or crisis situations outside its borders, in order to preserve or create a balance of power favorable to its interests."[23] More recently, a Chinese state-owned newspaper warned, "China won't come to North Korea's aid if it launches missiles threatening U.S. soil and there is retaliation, . . . but it would intervene if Washington strikes first."[24] Alternatively, in 2014, Xi Jinping also gave approval for peaceful Korean unification: "Xi reaffirmed his full support for Park's push for unification, expressing his hope that 'the two Koreas achieve unification in an independent and peaceful way by carrying out a process of reconciliation and cooperation.'"[25] Of course, Xi might have said this in an effort to gain support from the two Koreas while expecting that a peaceful unification would never happen.

Consistent with Xi's concerns about North Korean chaos, Chinese security experts talk about intervening in North Korea to deal with various threats posed to China.[26] The principal threat they discuss is North Korean refugees crossing into China, because these refugees would likely have difficulty finding jobs and could therefore rebel and destabilize the large ethnic Korean minority in northern China.[27] China would probably also want to secure North Korean WMD

---

[23] Michael Auslin, "In Search of the Xi Doctrine," War on the Rocks, May 30, 2016.

[24] Simon Denyer and Amanda Erickson, "Beijing Warns Pyongyang: You're on Your Own If You Go After the United States," *Washington Post*, August 11, 2017.

[25] "Korea-China Summit," editorial, *Korea Herald*, March 25, 2014.

[26] Bonnie Glaser, Scott Snyder, and John S. Park, *Keeping an Eye on an Unruly Neighbor: Chinese Views of Economic Reform and Stability in North Korea*, Washington, D.C.: Center for Strategic and International Studies and U.S. Institute of Peace, January 3, 2008.

[27] For example, according to a *Daily Mail* article, "The Chinese army has reportedly deployed 150,000 troops to the North Korean border to prepare for pre-emptive attacks after the United States dropped airstrikes on Syria. . . . The troops have been dispatched to handle North Korean refugees and 'unforeseen circumstances,' such as the prospect of preemptive attacks on North Korea" (Kelly Mclaughlin, "China 'Deploys 150,000 Troops to Deal with Possible North Korean Refugees over Fears Trump May Strike Kim Jong-un Following Missile Attack on Syria,'" *Daily Mail*, April 10, 2017).

and related delivery systems,[28] many of which are much closer to the Chinese border than to the ROK border. After all, Chinese leaders know about the 1993 meeting held by Kim Il-sung, described earlier, in which Kim Jong-il threatened to destroy the Earth. China could well be targeted by North Korean WMD in circumstances that North Korea feels would be adverse to the future of the regime. According to a report by Bonnie Glaser, Scott Snyder, and John S. Park,

> If deemed necessary, PLA troops would be dispatched into North Korea. China's strong preference is to receive formal authorization and coordinate closely with the [United Nations] in such an endeavor. However, if the international community did not react in a timely manner as the internal order in North Korea deteriorated rapidly, China would seek to take the initiative in restoring stability.[29]

There have also been several reports in recent years of Chinese military training to intervene in North Korea.[30]

From a ROK perspective, any Chinese intervention in North Korea would be an invasion of Korea because the ROK constitution

---

[28] As Glaser, Snyder, and Park (2008, p. 19) notes,

> According to PLA [People's Liberation Army] researchers, contingency plans are in place for the PLA to perform three possible missions in the DPRK. These include: 1) humanitarian missions such as assisting refugees or providing help after a natural disaster; 2) peacekeeping or "order keeping" missions such as serving as civil police; and 3) "environmental control" measures to clean up nuclear contamination resulting from a strike on North Korean nuclear facilities near the Sino-DPRK border and to secure nuclear weapons and fissile materials.

[29] Glaser, Snyder, and Park, 2008, p. 19.

[30] For example, a *Newsweek* article notes,

> Chinese interests in a Korea contingency have expanded beyond concerns about a refugee spillover to include nuclear security. Chinese military capabilities have improved greatly over the past 10 years, and the missions the People's Liberation Army (PLA) may be involved in have expanded in tandem. Training, equipment, exercises and aspects of the reorganization suggest contingency plans are likely in place for a mission to secure North Korean nuclear weapons and fissile material. (Oriana Skylar Mastro, "Will China Invade North Korea and Take Its Nuclear Facilities?" *Newsweek*, September 14, 2017)

defines all of North Korea as part of Korea.[31] If China were thus to "invade Korea," many South Koreans would feel compelled to launch a counterattack, potentially causing a ROK war with China. Even with U.S. assistance for South Korea, such a conflict could cause far more damage to both North and South Korea than a conflict between the North and South alone could. In an extreme case, a war between South Korea and China could lead to China conquering all of Korea. But at the very least, the decimation caused would make unification difficult to achieve and even more difficult to sustain.

Even if war between South Korea and China were avoided, Chinese forces intervening in North Korea could likely occupy a substantial portion of the country before they came in contact with ROK or U.S. forces. Although China might not prefer to occupy this territory in the long term, it might feel compelled to do so until South Korea could stabilize the areas of North Korea below the Chinese occupation. For the reasons discussed, South Korea might have difficulty stabilizing the parts of North Korea that it initially occupies, potentially convincing China that it needs to perform a longer-term occupation that frustrates full Korean unification. This would, in turn, allow South Korea to unify only the initially occupied parts of North Korea. Alternatively, if South Korea is successful at stabilizing parts of North Korea, China might still make a series of demands before withdrawing from North Korean territory. These demands might include allowing Chinese firms to retain North Korean companies and resources that they have taken over, and these firms are assets that South Korea has hoped would help pay for Korean unification. South Korea should expect and prepare for Chinese demands.

---

[31] Republic of Korea, "Constitution of the Republic of Korea," National Assembly of the Republic of Korea, October 29, 1987, Article 3.

Conclusions

ROK, U.S., and DPRK military forces would have a role in any form of unification. This is most clear in the paths involving warfare or in a DPRK government collapse after which ROK and U.S. forces intervene. Furthermore, in the paths based on peaceful unification, military forces will still have a role in demilitarizing the opposing military and security forces and stabilizing the unification. But as noted, South Korea (and the United States) might decide not to demilitarize all North Korean forces, instead keeping some in uniform to assist with stabilization. Doing so might also prevent these North Korean personnel from defecting to insurgent organizations.[32] Inadequate ROK and U.S. military capabilities could cause unification to fail, regardless of the unification path followed.

Table 2.2 provides an initial summary of how these seven challenges could prevent a full unification or make it impossible to sustain any initial unification. The table suggests that full Korean unification might be difficult to achieve—and even more difficult to sustain.

---

[32] The disbanding of the Iraqi Army in 2003 provided support for the Iraqi insurgencies, and that should be avoided to the extent possible in a Korean unification.

Table 2.2
How the Challenges Could Affect Achieving or Sustaining Korean Unification

Challenge	Achieving Full Unification	Sustaining Unification
North Korean WMD	• Cause irrevocable damage and destruction of resources needed to begin economic recovery in the North	• Leave too few ROK forces for stabilization • Lead to revenge actions by ROK citizens against North Koreans
Other North Korean military forces	• Significantly resist ROK and U.S. advances • Make establishing control over all of North Korea impossible	• Form a resistance beyond the ROK ability to stabilize, even after North Korean demilitarization
Reduced size and readiness of the active-duty ROK Army	• Leave the force too small and ill-prepared to defeat the DPRK military, impose unification, or stop North Korean conquest of South Korea • Lead to a North Korea–led unification	• Leave the force too small and not prepared to stabilize North Korea after conflict with the North
Inadequate training for ROK Army reserve forces	• Leave the reserves inadequately prepared to meaningfully augment the active Army	• Result in inadequate ROK military cohesion for stabilizing North Korea
Slow deployment of U.S. forces to Korea	• Cause U.S. forces to arrive after ROK forces are damaged, leaving South Korea vulnerable or delaying advances into the North	• Leave reconstituted ROK forces inadequately prepared to stabilize North Korea
Inadequate ROK planning for a replacement government in North Korea	• Stall the ROK and U.S. advance into the North	• Create too big a North Korean resistance to stabilize
Chinese intervention in North Korea	• Allow China to secure some or much of North Korea before South Korea and the United States can, preventing full unification	• Give China reason to conclude that it must hold North Korean territory to achieve stability

CHAPTER THREE

Unification Paths Resulting from War

For more than 60 years, the United States has helped prepare South Korea to defend itself against a major war initiated by North Korea. And although North Korea was pursuing a charm offensive in early 2018 that presents the regime as a partner for peace on the Korean Peninsula, North Korea continues to build up cyber capabilities that attack South Korea and other countries, as well as military capabilities that could be used to attack South Korea. North Korea has apparently been deterred from military attacks because it would have little likelihood of succeeding at an acceptable cost, it would face high risks if WMD were employed, and the end result would probably be the destruction of the DPRK regime. After all, in a major conflict, South Korea could suffer hundreds of thousands to millions of casualties, coupled with substantial damage to the ROK economy. So, how could the ROK government allow the North Korean regime to survive and be able to attack yet again at a later time? Thus, if North Korea attempts to conquer South Korea, South Korea could seek the destruction of the North Korean regime and its military, as well as unification of Korea under ROK control.[1] But ongoing ROK military force reductions might make these objectives infeasible.

---

[1] As suggested in the appendix, a future war in Korea could cost millions of lives. The ROK government would likely not want to leave a defeated but surviving DPRK regime and military in existence that could build up its military capabilities for a few years and then try again to conquer South Korea. The U.S. and allied objectives against Germany and Japan were similar in World War II.

This chapter examines three possible paths to Korean unification in the aftermath of a major war on the Korean Peninsula. The first path is *War: DPRK conquest* of South Korea and involves a North Korean victory over South Korea, leading to a DPRK-led unification. The *War: ROK-U.S. conquest—costly victory, contested peace* path involves ROK and U.S. forces defeating North Korea and achieving a ROK-led unification using current plans, which would result in a costly victory and a contested peace. The *War: Improved ROK-U.S. conquest—reduced costs of victory, improved postconflict stability* path involves South Korea and the United States taking actions starting now to shape the potential course of the events in a possible future war, setting conditions that would lead to a more favorable outcome than the baseline *War: ROK-U.S. conquest—costly victory, contested peace* path.

The initial parts of each path could be very similar, but as with all major wars, there would be substantial uncertainty in the course of events, including the nature and sequence of attacks and eventually whether North Korea or South Korea obtains dominance in the conflict and is able to drive toward its preferred form of unification.

Initiating Events for War-Related Paths to Unification

The war-related paths to unification could be initiated in at least three different manners.[2] In the first (initiating option A), North Korea would seek conquest of South Korea from the beginning of a conflict, and eventually the North (*DPRK conquest* path) or the South (*ROK-U.S. conquest—costly victory, contested peace* path) would prevail. In the second (initiating option B), the United States would act on the threats it has made to carry out limited military attacks on North Korea to rein in the DPRK nuclear weapon program. In the third (initiating

---

[2]   While conceptually it would be possible for South Korea to initiate the conflict by invading the North, RAND has done considerable modeling of that path in recent years and found that it would be a disaster for South Korea because of the attrition that DPRK artillery and maneuver forces, likely using at least chemical weapons, could impose on ROK forces, which would have to advance in the open to invade the North. Therefore, such an initiating event is not considered in this report.

option C), North Korea would once again carry out the kind of limited military attacks it has made over the years on South Korea. In either initiating option B or C, the adversary responses could be very limited,[3] preventing escalation to a major conflict and thus not leading to anything close to Korean unification (and thus outside the scope of this report). In all of these initiating options, North Korea would be expected to launch major cyberattacks to disrupt ROK infrastructure, ROK society, and the ROK military. If the initiating events lead to major escalation, a full war could develop and thereby establish the potential for any of the three war-related paths to unification.

Initiating Option A: North Korean Invasion of South Korea

For decades, the North Korean military has prepared and trained for conquering South Korea. North Korea initially attempted to execute this plan in the Korean War of 1950 to 1953. That war resulted in serious attrition to the military forces on both sides, leaving both with inadequate military capability to conquer the other side and aborting the unification desired by both. Since that time, North Korean forces have prepared for a second invasion attempt. Recognizing the evolution of warfare over the years, North Korea began building substantial armored forces in the 1970s. But in the 1980s, North Korea adjusted its approach by retaining substantial infantry forces, limiting the amount of armor it built, and emphasizing the use of artillery and ballistic missiles, allowing for the delivery of WMD. North Korea apparently hopes to break U.S. and ROK military force cohesion and advance rapidly to conquer South Korea before substantial U.S. forces could reach the Korean Peninsula.

Initiating Option B: Limited U.S. Military Attacks on North Korea

North Korea's nuclear weapon capability, coupled with the intercontinental ballistic missiles (ICBMs) it is developing for nuclear weapon delivery, is increasingly perceived by the United States as an existential threat. The U.S. government recognizes that even a few adversary

---

[3] This has been the historical pattern with North Korean military attacks, such as those of 2009 and 2010.

nuclear weapons detonated on major U.S. cities could cause damage to the U.S. population and economy well beyond the physical nuclear weapon effects, leading to chaos and major disruption of the complex interactions required by a major modern economy and society.

As a result, President Trump has threatened to carry out "limited" attacks against DPRK nuclear weapon and missile facilities to destroy or at least seriously disrupt North Korea's nuclear weapon program and developing ICBMs. The United States would likely perform such attacks at least initially with conventional forces, hoping to severely degrade North Korea's nuclear weapon capabilities without causing major collateral damage or justifying serious escalation. But because the DPRK assets the United States would be targeting are viewed by North Korea as critical to the regime's survival, it is entirely possible that North Korea would respond to these attacks with a major escalation, potentially including the use of nuclear weapons or other WMD.

Alternatively, North Korea's initial response could be relatively limited, perhaps as a result of U.S. efforts to damage North Korean command, control, and communications networks. In such a case, the United States might decide to execute further counterforce and counterleadership attacks, hoping to disable the North Korean nuclear weapon threat. In turn, North Korea could decide to significantly escalate to a major war, especially if it feared the potential consequences of losing the conflict and decided to "destroy the Earth," as discussed in Chapter Two.

Therefore, a limited U.S. military attack on North Korea could lead to a major war in Korea in a variety of ways. Alternatively, it could remain a limited conflict, not leading to unification; such a case would be outside the scope of this report.

Initiating Option C: Limited North Korean Military Attacks on South Korea

Since the end of the Korean War, North Korea has carried out a series of limited military attacks against South Korea. In recent years, these provocations have included sinking a ROK warship and shelling a ROK island in 2010, as well as executing a landmine attack in the

DMZ in 2015. South Korea has become very sensitive to such attacks and has adopted a strategy of overwhelming response.

Such a ROK response risks North Korean escalation, against which South Korea could also escalate. Thus, an escalation spiral up to a major war could follow a North Korean limited military attack. On the other hand, ROK threats of strong retaliation have likely deterred many North Korean provocations. It is hard to determine the relative likelihood of such a spiral, but it is certainly possible and, in some cases, might be probable. Escalation spirals appear to be less likely if the response is prompt and local.

The War-Related Paths: The Early Stages of Warfare

Following any of the potential initiating options, a major war between North Korea and South Korea (with help from the United States) could develop. Because South Korea and the United States seek to avoid a major war, the decision to cross that threshold would likely be made by North Korea.

A major war in Korea could, in theory, proceed in at least three different ways, all of which would likely involve early and significant use of cyberattacks. First, North Korea could invade South Korea, consistent with initiating option A, but it could also be part of the escalation resulting from initiating options B and C. If North Korea were to invade, its ground forces would cross the DMZ with the support of a massive artillery barrage on the ROK defensive positions and other targets in the South. It would also be accompanied by major infiltrations of DPRK special forces that would attack deeper targets in the South, such as airfields; ports; and command, control, communications, computers, and intelligence assets. These attacks also would be accompanied by standoff missile, air, and naval attacks against South Korea. Second, South Korea could invade the North, but this is unlikely unless there is a collapse of the North Korean government, which would be the *DPRK collapse: Intervention* path discussed in the next chapter. Third, both sides could keep their conventional ground forces in place along the DMZ and use only standoff fires, including artillery and missile fire; air attacks; and special forces. The use of WMD by North Korea and nuclear responses by the United States might make these standoff attacks nearly as damaging as a full North Korean ground force invasion.

In its transition to a major war, North Korea would face a major quandary: How would it use its missile-based WMD? The North Korean quandary is driven by South Korea developing its own counterforce capability over the past nearly ten years, which it refers to as the *kill chain*.[4] The kill chain is a combination of South Korean ballistic and cruise missiles, fighter aircraft, and armed unmanned aerial vehicles (drones) being developed to allow South Korea the option for "damage-limiting" attacks against DPRK nuclear weapons, other WMD programs, and missile delivery systems. But because even a modest amount of WMD attacks on South Korea could cause major damage, South Korea plans to use the kill chain preemptively or possibly even preventively to stop North Korean attacks before they can be executed in a major war. South Korea consistently refers to the preemptive use of the kill chain.[5]

North Korea has responded that it can preempt the ROK preemption, which, if it were to do so, would probably involve the use of

---

[4] In 2012, the United States and South Korea announced that

> The two nations will establish a "kill chain" to detect, track and target North Korean missiles, and a Korean Air and Missile Defense System aimed at intercepting incoming missiles. If the kill chain is established, it would take less than 30 minutes to target a missile after detection, according to officials. (Kim Eun-jung, "S. Korea, U.S. Agree to Set N. Korean Nuclear Deterrence Policy by 2014," Yonhap News Agency, October 25, 2012)

[5] For example, according to a 2013 news article,

> The Joint Chiefs of Staff is currently working on the proactive deterrence strategy, which will include preemptive strikes. Before the National Assembly on March 6, Jung Seung-jo, chairman of the joint chiefs of staff, made clear that preemptive strikes on the North's nuclear facilities are a matter of exercising the right of self-defense and Seoul does not require Washington's consent to make them. With the new strategy, the South intends to launch preemptive strikes within 30 minutes of detecting signs of an imminent attack by the North using weapons of mass destruction including nuclear weapons. The backbone of the strategy will be the establishment of a so-called "Kill Chain." (Ser Myo-ja, "Park Tells Military to Strike Back If Attacked," *Korea JoongAng Daily*, April 2, 2013)

nuclear weapons and other WMD from the very beginning of a major conflict, whether an invasion or standoff attacks are the focus of the military operations. This would likely make a North Korean invasion of South Korea or other early action in a major war very different from what has historically been expected: From the very beginning of its attack, North Korea could use WMD to cause major damage to the militaries and societies of South Korea and the United States.

This quandary is further complicated by the subsequent development of South Korea's Korean Massive Punishment and Retaliation strategy. According to this strategy,

> Every Pyongyang district, particularly where the North Korean leadership is possibly hidden, will be completely destroyed by ballistic missiles and high-explosive shells as soon as the North shows any signs of using a nuclear weapon. In other words, the North's capital city will be reduced to ashes and removed from the map.[6]

Thus, the DPRK regime's survival would be held in jeopardy from the beginning of any major conflict. That jeopardy could cause North Korea to execute its "destroy the Earth" threat, potentially leading to massive damage early in any major war.

If North Korea invades South Korea, massive damage would result on both sides from attacks by conventional weapons, and more-significant damage would result from the use of any North Korean WMD or from U.S. nuclear retaliation. The major issue will become whether the ROK forces (including especially ground and air forces) would be able to stop the North Korean advance. If they cannot, the *War: DPRK conquest* path would develop. But if the ROK forces can stop the advance and have sufficient force and will to counterattack and advance into North Korea, the *War: ROK-U.S. conquest—costly victory, contested peace* path would develop. Otherwise, a variant of the current armistice could well develop over time, without unification.

---

[6] "S. Korea Unveils Plan to Raze Pyongyang in Case of Signs of Nuclear Attack," Yonhap News Agency, September 11, 2016.

War: DPRK Conquest Path

Should conflict between North Korea and South Korea develop into major war following any of these initiating options, the DPRK regime's survival objective (its primary goal) would almost certainly broaden to include the defeat of the ROK and U.S. militaries, destruction of the ROK government, and conquest of South Korea that leads to a DPRK-controlled unification. But could North Korea really conquer South Korea?[7]

In recent years, U.S. military commanders in Korea have significantly discredited the possibility of a North Korean conquest of South Korea. In 2005, the U.S. commander in Korea, GEN Leon LaPorte, said to the Senate Armed Services Committee, "We remain confident in our ability to deter North Korean aggression and, if necessary, capable of defeating aggression against the Republic of Korea."[8] U.S. and ROK forces had, by that time, obtained a substantial degree of conventional qualitative superiority over DPRK forces. It was perceived that, in a purely conventional conflict, the qualitative superiority of U.S. and ROK forces would overcome North Korea's quantitative superiority. As of 2009, GEN Walter Sharp, the then-commander of U.S. military forces in South Korea, said he was certain that U.S. forces could "defend against any threat from communist North Korea."[9] And more recently, Marine Corps Gen. Joseph Dunford, the chairman of the Joint Chiefs of Staff, and Army GEN Vincent K. Brooks, the allied commander in Korea, said that the capabilities already in South Korea were "enough to defend against a strike from Kim Jong Un, the North Korean dictator."[10]

---

[7]   See, for example, Michael Peck, "North Korea Plans to Defeat the U.S. Army in a War. Here's How," *National Interest*, January 12, 2018.

[8]   U.S. Senate, *Fiscal Year 2006 Defense Department Budget: Hearing of the Senate Armed Services Committee*, Washington, D.C., U.S. Government Printing Office, March 8, 2005.

[9]   "U.S. General Says Forces Ready to Counter N.Korean Attack," *Chosunilbo*, July 15, 2009.

[10] Jim Garamone, "US-South Korean Alliance Ready to Defend Against North Korean Threat, Top Generals Say," DoD News, Defense Media Activity, August 15, 2017.

It is not clear that these military assessments adequately accounted for North Korea's WMD and other asymmetric threats, such as the North's special forces and cyber capabilities. But North Korea has been deterred from starting a major conflict, partly by recognizing that U.S. nuclear weapons would likely be used to respond to its WMD use. Still, if North Korea finds itself in a major war, ROK and U.S. conventional superiority, coupled with the Kill Chain and the Korean Massive Punishment and Retaliation strategy, would likely force North Korea into using WMD—potentially early and extensively.

The United States announced its planned response to North Korean nuclear weapon use in its 2018 Nuclear Posture Review, which stated,

> Our deterrence strategy for North Korea makes clear that any North Korean nuclear attack against the United States or its allies and partners is unacceptable and will result in the end of that regime. There is no scenario in which the Kim regime could employ nuclear weapons and survive.[11]

Similarly, North Korean use of chemical and biological weapons against U.S. military facilities in South Korea would likely kill or seriously injure large numbers of military and civilian personnel of the U.S. Department of Defense and military dependents. Americans also would be killed in any attacks on ROK cities. The resulting U.S. retaliation against North Korea would likely leave North Korea in no better shape and potentially much worse shape than South Korea.

Assuming that North Korean ground forces could advance as far as Seoul, it would be a largely Pyrrhic victory. The Kim family would likely be dead, both Koreas would be decimated, and any significantly sized DPRK ground force would be destroyed by U.S. and ROK air attacks if it continued as far south as Pyeongtaek (the location of Camp Humphreys, at least 100 km from the DMZ), let alone Pusan (400 km from the DMZ). Thus, North Korea would be unlikely to achieve a meaningful Korean unification under its control—not

---

[11] Office of the Secretary of Defense, *2018 Nuclear Posture Review*, Washington, D.C.: U.S. Department of Defense, February 2018, p. 33.

unlike the situation that led to the armistice in 1953. But that armistice came only at the end of three years of brutal and bloody warfare; a future major war between North Korea and South Korea could create such a stalemate within days because of the increasing deadliness of the weapons involved.

War: ROK-U.S. Conquest—Costly Victory, Contested Peace Path

Following any of the potential initiating conditions, a major war with North Korea could develop in which the uncertainties of combat lead to ROK and U.S. dominance over parts of North Korea as the conflict proceeds; this is the alternative to North Korean dominance as examined in the *War: DPRK conquest* path. As ROK and U.S. forces enter DPRK territory, they would likely want to destroy the DPRK regime, neutralize the military, capture all DPRK territory, and unify Korea under ROK control. But even if North Korea loses the conflict, the DPRK regime could do immense damage to the South. The thousands of artillery systems along the DMZ alone could decimate Seoul and other forward city areas, damaging both the people and the economy of South Korea.[12] The addition of North Korean nuclear and other WMD attacks on the South could result in the loss or incapacitation of hundreds of thousands or perhaps millions of South Koreans and cause substantial damage to the ROK economy, much like in the *War: DPRK conquest* path. And ROK and U.S. counterattacks and retaliation would almost certainly cause even more damage in North Korea. The level of damage on both sides would probably be so serious that ROK and U.S. forces might not be able to capture all of North

---

[12] Some experts are skeptical about North Korea's artillery threat; see, for example, Prakash Menon and P. R. Shankar, "Could North Korea Destroy Seoul With Its Artillery Guns?" *National Interest*, May 25, 2018. Still, if North Korea's initial artillery barrage is synchronized, then before South Korea could fire its first reply, the North could deliver well over 5,000 shells and artillery rockets to central Seoul and much more to Seoul's northern suburbs—enough to cause substantial damage and serious psychological scars. If this artillery fire included chemical weapons, as might be expected, the human casualties could be far worse.

Korea's territory, especially if China intervenes in the North and if the ROK Army in peacetime has been reduced to fewer than 400,000 active-duty personnel.

A ROK-led unification of Korea resulting from a war could be only a partial unification. Moreover, the surviving ROK and U.S. forces might prove insufficient to establish stability in all of the North Korean territory that would likely be captured. If so, the ROK-U.S. alliance should expect a North Korean insurgency that frequently disrupts the unification and stability operations. If South Korea and the United States are able to capture Pyongyang, the ROK government would likely plan to purge the DPRK government personnel and possibly even technocrats, fearing their disloyalty. And even if some of these personnel and technocrats are not outright opponents, they nonetheless could try to subtly disrupt the unification. Thus, the ROK government would expect to take control of the DPRK government, but attrition to the ROK individuals chosen to replace North Korean leaders might undermine the ability to do so and create a hostile group of former elites that could assist or lead an insurgency against a ROK-controlled government in the North.[13] As a result, the government organizations in North Korea would face potentially significant opposition and might not function well for many years, lacking needed familiarity and expertise, which would lead to significant disaffection among the North Korean people.

Finally, the damage caused by such a major war would potentially compel Chinese intervention in the North. Even if North Korea were to execute WMD attacks against Chinese forces, the Chinese could likely reach Pyongyang, a potential objective, before ROK and U.S. forces would be able to, leaving China in control of much of North Korea. This situation could lead to conflicts between ROK forces and Chinese troops, especially if the ROK military seeks to expel Chinese

---

[13] South Korea plans to replace the North Korean provincial, city, and town leaders in this kind of scenario, partly by bringing in personnel from the ROK Commission of the Five Provinces, which consists of the ROK-appointed governors, mayors, and other government officials who plan to take over North Korean positions. See Alastair Gale and Kwanwoo Jun, "South Korea's Governors of Northern Provinces Don't—and Never Will—Govern," *Wall Street Journal*, March 17, 2014.

forces from at least Pyongyang and the surrounding area. In the circumstances of this unification path, it is very unlikely that ROK forces would be able to push the Chinese forces back.

If South Korea also has difficulty stabilizing the part of North Korea that it initially occupies, China would be reluctant to withdraw its forces from North Korea if it expects chaos and anarchy to erupt in the areas that Chinese forces abandon.

War: Improved ROK-U.S. Conquest—Reduced Costs of Victory, Improved Postconflict Stability Path

The likely outcomes of the *War: DPRK conquest* path and the *War: ROK-U.S. conquest—costly victory, contested peace* path would not be acceptable to South Korea or the United States.[14] To make the outcomes of a major war in Korea more acceptable, South Korea and the United States must make two important changes to current conditions, and these changes create the *War: Improved ROK-U.S. conquest—reduced costs of victory, improved postconflict stability* path. First, South Korea and the United States must make an effort to co-opt North Korean elites by convincing them that they will have a good life after a ROK-led unification. Doing so requires, in part, a change in perspective: The unification should be led by South Korea and should not be characterized as a conquest of North Korea. Second, South Korea and the United States must prepare for warfighting in a WMD environment, including continued fighting to stabilize North Korea after the major warfighting has concluded.

Co-Opting North Korean Elites

One of the challenges noted in Chapter Two is inadequate ROK planning for a replacement government in North Korea—in other words, what to do with North Korean elites. North Korean indoctrination

---

[14] Of course, it would be far better for South Korea and the United States to establish a condition of true and lasting peace on the Korean Peninsula rather than fighting a war, but such a condition is outside the scope of this chapter. It will be addressed in Chapter Five.

tells the elites how bad their lives will become in a ROK-controlled unification, and South Korea and the United States have done little to convince them otherwise. As I noted in a 2017 report,

> At unification, according to one North Korean propaganda statement, the United States and the ROK will "exterminate the core class families first." Even if the broader ranks of the core class (the estimated 4.4 million adult elites) discount such regime statements as extreme and exaggerated, senior elites (perhaps several hundred thousand) likely will worry about their postunification fates. More broadly, North Korean elites—the core class—expect that unification led by the ROK would jeopardize their positions, safety, and security, potentially leading to their imprisonment or worse. These changes would affect not only the elites but also their families, giving them little potential for a good life, let alone the kind of privileged lives to which they have grown accustomed.[15]

South Korea and the United States need to address these various issues in ways that reassure North Korean elites, including telling many of them that they would be allowed to retain their government roles after unification. A transitional justice system will be required, but wholesale purging of Koreans in the North might contribute to resistance and instability, as well as a loss of tremendous local knowledge and capability. Appropriate approaches will be discussed in more detail in Chapter Five.

If North Korean elites think that they will have good lives after ROK-led unification, then it might be possible to co-opt them to some degree. South Korea and the United States would want cooperation from these elites to break the DPRK command and control system associated with using WMD against South Korea and its forces. Although Kim Jong-un has claimed that he has a button on his desk that he can push to launch his nuclear forces,[16] it seems more likely that Kim would need to order the commander of his Strategic Forces to launch nuclear

---

[15] Bennett, 2017, p. 9.

[16] Kim said, "the nuclear button is on my office desk all the time" (Kim Jong-un, 2018).

weapons, who would then need to order each missile base commander, who would then need to order each missile battalion commander, who would then need to order each missile battery commander to launch missiles carrying nuclear weapons. If that is the case, the DPRK chain of command could be broken by any one of these individuals not following the order to launch or by directing the launch of a nonnuclear weapon.[17] South Korea and the United States need to pursue psychological operations that inform DPRK commanders at each level about the future that is possible for them and their families if they do not use nuclear weapons, as well as the consequences for them personally if they do order such use.[18] These DPRK commanders will be members of elite families, many of which have become capitalists since the famine of the 1990s.[19] Their sympathies might be less with the regime and more with the future of their families and their enterprises. Thus, deterrence of nuclear weapon use can, in theory, be achieved at multiple command levels and not just by deterring Kim Jong-un.

The perceptions of North Korean elites will also be critical if South Korea and the United States attempt to stabilize DPRK territory. If North Korean elites are treated badly, removed from their positions and jobs, arrested for bribery or other offenses (in some cases), and left with meaningless lives, many of those elites could decide to participate in insurgent activities. South Korea and the United States currently appear to have few plans to prevent such outcomes, and they need to develop and resource appropriate continuity and stabilization plans, including a transitional justice plan that supports rather than challenges unification.

---

[17] Of course, the DPRK military has a political control system designed to assure that all orders are carried out. But in wartime, such a system could break down in a variety of ways, including South Korea or the United States co-opting both the military commander and his political control officer(s).

[18] South Korea probably needs a very public effort to create laws that would designate as a war criminal anyone launching a WMD attack against South Korea.

[19] Senior North Korean defectors tell me that, as markets began developing in North Korea in the late 1990s and early 2000s, many of the elite families concluded that they would be far better off participating in the market activity. Many did so, and they are largely responsible for the restaurants, shops, cell phones, and other signs of capitalism in North Korea.

In addition, South Korea and the United States should seek internal information about North Korea's WMD and related programs. South Korea could announce that it is prepared to offer a substantial reward to anyone in the DPRK Strategic Forces or the Second Economic Committee who defects to South Korea and can provide information on North Korean WMD and associated delivery systems.[20] The rewards would be based on the value of the information provided and could be quite large. For counterforce efforts to be successful, South Korea and the United States need to enhance their information on North Korean WMD threats, and appropriate defectors could help in this process.[21]

Warfighting in an Environment with Weapons of Mass Destruction
South Korea and the United States also need to better prepare for warfighting in a WMD environment. Co-opting North Korean elites could assist in this area by potentially deterring WMD use by some individual DPRK commanders, reducing the cohesion of DPRK forces, and developing more information to support targeting of the WMD program.

But there also needs to be considerable development of strategy, operations, and capability to survive and operate in a WMD environment. ROK and U.S. forces need to be adequately protected against North Korean WMD use. Such protections include acquiring adequate individual protective equipment; creating more blast and fallout shelters; dispersing military equipment; developing better intelligence and counterforce capabilities to reduce the WMD threat before it is launched; improving the capability to maneuver around WMD-contaminated areas, especially in the channelized terrain of North Korea; developing greater redundancy of command, control, communications, computers, and intelligence; and recruiting and training

---

[20] The North Korean Second Economic Committee is the North Korean organization running the North's military industry.

[21] Of, course, if South Korea and the U.S. advertise for defectors, the North Korean security services will probably send agents pretending to be defectors to mislead ROK authorities. There would need to be a significant vetting process put in place and plans for handling false defectors.

more forces to provide replacements in the event of attrition of ROK ground forces. ROK Army reserve divisions, in particular, need to be better prepared to perform a range of missions against North Korea, as discussed in Chapter Two.

As noted previously, the 2018 U.S. Nuclear Posture Review has stated that, if the DPRK regime employs nuclear weapons, it will not survive. If the regime uses nuclear weapons early in a conflict, South Korea and the United States need to be prepared to rapidly decapitate the DPRK regime, hoping to terminate or seriously reduce further WMD use.[22] Intelligence on the regime's, and especially Kim Jong-un's, location would be key to accomplishing this task. This information might be obtained by co-opting other senior DPRK leaders and obtaining information from them (1) before a conflict, seeking to learn about the locations Kim frequents, and (2) during a conflict, seeking information about his real-time location. South Korea and the United States would then need to be prepared to promptly attack such locations and employ bomb damage assessment to confirm or deny Kim's destruction so that further attacks could be mounted if needed.[23]

The efforts outlined for the *War: Improved ROK-U.S. conquest—reduced costs of victory, improved postconflict stability* path should have a positive effect on a ROK-led unification. Nevertheless, the potential outcomes of the *War: ROK-U.S. conquest—costly victory, contested peace* path, which describes a ROK-led unification under current circumstances, are so bad that it is not at all clear that even the improved path described in this section will offer an acceptable outcome. Therefore, some of the paths related to regime collapse or peace would be better options, if they are available.

---

[22] There is the risk that if Kim Jong-un is eliminated promptly, he could have designed a "fail deadly" nuclear release system that would launch all remaining North Korean nuclear weapons as soon as possible if control from Kim is lost. But given Kim's paranoia about threats to the control of his military, it seems unlikely that Kim would establish such a system.

[23] For deterrence purposes, Kim should be told that his WMD program requires South Korea and the United States to plan an early and thorough decapitation of his regime. Such statements might put more pressure on Kim to meet his commitments to abandon his nuclear weapons.

Conclusions on War-Related Paths to Unification

Each of the initiating events in these three paths has a moderate likelihood of happening today, although initiating events B and C have a low likelihood of leading to a major war. The *War: Improved ROK-U.S. conquest—reduced costs of victory, improved postconflict stability* path is unlikely today because of a failure of South Korea and the United States to adequately prepare for major warfare in Korea. But if those preparations are made, that path has a higher likelihood. It could still involve severe casualties and destruction, but the damage would not be nearly as bad as that of the other two war-related paths to unification. Across the challenges, the attrition levels (largely resulting from WMD use) and Chinese intervention appear to have the most-serious impacts.

Since the end of the Cold War, the United States has experienced conflicts with historically low attrition (i.e., casualty) levels. Even if a major war in Korea were limited to the use of conventional weapons, this historical experience will almost certainly be broken by much higher levels of military and civilian attrition in Korea. The addition of substantial quantities of WMD, and especially of nuclear weapons, could cause unprecedented attrition levels in a major war in Korea. The outcomes of such wars could thus be very different from what even experts have expected. In the *War: DPRK conquest* path, the attrition would likely be so serious as to prevent North Korea from being able to fully conquer South Korea (at least in any time-urgent manner), and a similar outcome would likely be true for the *War: ROK-U.S. conquest—costly victory, contested peace* path.

For all three paths, the damage done by WMD to both military forces and civilians on both sides would likely limit the initial unification to being partial at most. Although the most intense phases of combat would likely occur in the first days and weeks, it would probably take months to create an initial partial unification. And given the probable levels of damage, it is difficult to see how either side would be able to achieve a stabilized partial unification.

In the *War: ROK-U.S. conquest—costly victory, contested peace* path and the *War: Improved ROK-U.S. conquest—reduced costs of victory, improved postconflict stability* path, Chinese intervention would play a major role. China would likely intervene early to destroy North Korean WMD that might be used against China and to otherwise establish stability in its border area.[24] Given the approximate locations of DPRK missile bases, China would likely have an initial objective of advancing to the narrow neck of North Korea, between roughly the cities of Anju and Hamhung.[25] In doing so, China would likely make its own forces and perhaps Chinese territory subject to WMD attack, although not of the magnitude experienced by South Korea and North Korea. Given the projected devastation in both North Korea and South Korea, China might conclude that it has little choice but to advance beyond the narrow neck, potentially down to the area around Pyongyang (which would likely be significantly damaged by ROK and U.S. standoff attacks). Chinese forces would be expected to reach Pyongyang before ROK or U.S. forces, leaving South Korea with a very limited Korean unification, and then only if it is able to overcome the heavy concentration of DPRK military forces south of Pyongyang. Even with just this area, the limited ROK forces, having likely suffered serious attrition, might be challenged in attempting to stabilize the occupied areas. Depending on South Korea's success in stabilization and other factors, China might or might not decide to withdraw from at least some of the areas it occupies. China is more likely to withdraw in the conditions of the *War: Improved ROK-U.S. conquest—reduced costs of victory, improved postconflict stability* path than in the conditions of the *War: ROK-U.S. conquest—costly victory, contested peace* path.

---

[24] As noted earlier, "contingency plans are in place for the PLA . . . to clean up nuclear contamination resulting from a strike on North Korean nuclear facilities near the Sino-DPRK border *and to secure nuclear weapons and fissile materials*." (Glaser, Snyder, and Park, 2008, p. 19, emphasis added).

[25] On the missile base locations, see Joseph S. Bermudez, Jr., "Behind the Lines: North Korea's Ballistic Missile Units," *Jane's Intelligence Review*, June 14, 2011.

CHAPTER FOUR

Unification Paths Resulting from Regime Collapse

For decades, there has been speculation about a potential collapse of the DPRK regime. Kim Jong-un's paranoia about the survival of his regime is illustrated by his having his uncle and older half-brother killed and hundreds of senior elites purged since he became the leader of North Korea in December 2011. Nevertheless, Kim's paranoia does not confirm that his regime is unstable; rather, it raises the possibility that the regime has stabilized as a consequence of the purges. Of course, even if the Kim family regime is eliminated, some form of government in North Korea could survive.[1]

Thus, in this chapter, I examine two possible paths to unification associated with a failure of the DPRK regime without predicting how or when such a failure might occur. The *DPRK collapse: Intervention* path involves the classical concept of a DPRK regime failure leading to a DPRK government failure that forces South Korea to intervene militarily in the North and ultimately to unify as much of Korea as possible over a short period. The *DPRK collapse: Negotiation* path postulates a failure of the Kim family regime that leads to one or more factions taking control of the various parts of North Korea and establishing one or more successor governments. These factions are more inclined to establish relations with the outside world, recognizing the

---

[1] Experts do not know whether a central government can survive if Kim Jong-un is suddenly no longer available. Instead, the DPRK government may fracture into factions, with some parts of the country lacking government authority. This highlights the importance of prompt ROK and U.S. action of some form (intervening or trying to sustain a central government) soon after recognizing the loss of Kim Jong-un.

economic benefits, and eventually agree to work with the ROK government toward a Korean unification.

DPRK Collapse: Intervention Path

Under this path, Kim Jong-un dies and there is no Kim family successor. As a result, the DPRK government fails, and the country breaks into rival factions that initiate a partial civil war. The DPRK military divides into these various factions, and control of the country's WMD and delivery systems are also split among the factions. Then, the factions start using some WMD in the civil war.[2] In previous work, I have analyzed the ROK and U.S. military forces that would be needed to make a unification of Korea work under less-serious collapse circumstances.[3]

As this unification path continues, South Korea, fearing a spillover of the civil war and Chinese intervention, launches its forces into North Korea and is supported by the United States.[4] The United States begins a full force deployment to Korea to support the ROK effort, but the initial effort on the ground depends almost entirely on ROK forces, given the limited peacetime presence of U.S. ground forces in South Korea. South Korea initially makes good progress into the North because North Korea's DMZ defense has lost a degree of coherence in the civil war. But the ROK (and Chinese) intervention gradually unifies the North Korean factions, allowing them to pose a more stout

---

[2]   Although South Korea would likely be cautious in intervening, by waiting very long to do so, it accepts many risks, including a deteriorating situation in the North and potential Chinese intervention.

[3]   Bennett and Lind, 2011. This paper identifies the minimum military requirements, assuming that the largest faction essentially surrenders authority to South Korea. The analysis was designed to put a lower bound on the forces needed. If the circumstances are less favorable, larger forces would be required.

[4]   Indeed, fearing the costs of intervention and of trying to absorb North Korea, South Korea might refrain from intervention until Chinese forces enter North Korea. But as argued in Chapter Two, even if South Korea is initially cautious, a Chinese military intervention in North Korea is highly likely to stimulate a ROK military intervention in the North to prevent Chinese domination of all of North Korea.

defense supported by the use of WMD against ROK and U.S. (and Chinese) forces. The WMD use starts initially as chemical artillery supporting the DPRK battlefield defense but eventually transitions to broader WMD use, including of nuclear weapons.

If China does not intervene initially, it does so soon thereafter, facing declining security across its border with North Korea. China seeks to stabilize its border area and avoid spillover of the DPRK civil war. In addition, China acts to eliminate North Korea's WMD. As in the *War: ROK-U.S. conquest—costly victory, contested peace* path, China would likely have an initial objective of advancing to the narrow neck of North Korea (between Anju and Hamhung) to secure most of the DPRK missile bases and other WMD sites in the northern part of North Korea.[5] In this path, as well, China could then decide to advance to Pyongyang. Most of the forces of the Chinese Northern Theater Command and perhaps some forces from its Central Theater Command would need to be committed to achieve China's objectives. Depending on how far the Chinese forces advance, they would likely perform most of the seizing and elimination of North Korea's WMD.

DPRK forces are much more numerous and better prepared along the DMZ than along the Chinese border. Therefore, in this scenario, DPRK forces likely have more effect in slowing ROK forces than Chinese forces as they advance, depending on the cohesion of the DPRK forces along the DMZ. Both sides presumably are trying to reach and control Pyongyang. Because the distance by road from the Chinese border to Pyongyang is very similar to the distance by road from the DMZ to Pyongyang (both are just over 200 km), and because Chinese forces would likely face lighter DPRK military opposition along the way, they have an initial advantage in trying to reach Pyongyang. But the uncertainties of warfare could allow either side to reach the capital first.[6]

---

[5]   On the missile base locations, see Bermudez, 2011.

[6]   Although the Chinese forces have sufficient combat power and would face far less DPRK opposition, they have not had to perform this kind of ground operation in recent years and thus would predictably suffer some challenges. For example, despite recent exercises to improve logistical support capabilities, Chinese forces likely would face logistical challenges. As a result, there is great uncertainty in the actual rate of Chinese advance and China's ability to control a broader area beyond the roads along which it would focus its penetrations.

There is a significant risk that, as ROK and Chinese forces approach each other, conflict between South Korea and China could develop. After all, as noted in Chapter Two, South Korea could well treat a Chinese intervention in the North as an invasion of Korea.[7] And even if South Korea decided not to try to expel the Chinese forces, accidents could develop between the two forces and escalate to greater conflict. A major war between South Korea and China would be a losing proposition for South Korea and thus ought to be avoided.

Depending on the level of North Korea's escalation to WMD use, the ROK forces aided by U.S. forces might be able to stabilize the area of North Korea south of Pyongyang, which is more rural and less heavily populated. But ROK forces moving into the Pyongyang and Wonsan region, if they do, would find stabilization to be a much more daunting task: The ROK forces in these areas could face a serious DPRK insurgency. As noted in Chapter Three, a ROK government plan to replace much of the DPRK government[8] would likely make the local governments in North Korea relatively ineffective for several years, having lost needed familiarity and expertise. What is known of the ROK government's current plans suggests that many North Korean elites would be quickly alienated, increasing the severity of the insurgency that South Korea would likely face—and thus increasing the difficulty of stabilizing the unification.

The bottom line of the *DPRK collapse: Intervention* path is that it leads to only a partial initial unification of Korea, given a Chinese intervention. The Chinese government likely would not plan to occupy and own parts of the North going forward, so the ROK government needs to plan for negotiations with the Chinese government to secure the withdrawal of its forces.

After South Korea is able to achieve a partial unification in this path, it would then face serious questions about sustaining the unifi-

---

[7]   Article 3 of the ROK constitution says, "The territory of the Republic of Korea shall consist of the Korean peninsula and its adjacent islands." Thus, any Chinese intervention into North Korea would be perceived by many in South Korea as an invasion of South Korea. See Republic of Korea, 1987, Article 3.

[8]   See Gale and Jun, 2014.

cation. In particular, a ROK occupation of Pyongyang, likely a major ROK goal, would force South Korea into dealing with a potentially serious insurgency, especially if South Korea expels North Korean elites from their positions of authority, leaving them without jobs. A particular threat would be the DPRK security forces and related military forces that are heavily based in Pyongyang and would likely be opposed to a ROK occupation that would treat many of them as criminals. South Korea might have difficulty suppressing the resulting insurgency. South Korea would likely do better stabilizing the North if it plans for good treatment of North Korea's elites and has carried out significant psychological operations well ahead of the regime's collapse to reassure the elites of a good future under a ROK-led unification. Waiting to take such action after regime collapse would likely be too late to gain that trust; the North Koreans would have difficulty believing the sincerity of ROK changes in policy and could seek to disrupt the unification.

DPRK Collapse: Negotiation Path

In this path, the Kim family no longer controls North Korea. Most likely, this has happened because North Korean elites have overthrown and perhaps even killed Kim Jong-un, although Kim could also suffer a health incident that would incapacitate or kill him. There is no apparent member of the Kim family who could take Kim Jong-un's place.

After the Kim family regime collapses, one or more successor factions could develop that would presumably be more inclined to work with the ROK government if its policies are favorable because the faction leaders would likely be capitalists.[9] Of course, the ROK govern-

---

[9]   The faction leaders would almost certainly be corrupt communist officials, consistent with the culture of DPRK leadership. But according to North Korean defectors, the vast majority of these personnel have also become involved in North Korean markets, meeting the definition of a capitalist: "a person who uses their wealth to invest in trade and industry for profit in accordance with the principles of capitalism" (Oxford University Press, English: Oxford Living Dictionaries, web tool, undated).

ment would not necessarily know which factions and leaders it could trust and could face significant difficulties in working with some of the factions. Indeed, from a ROK perspective, it would be best to try to work with the DPRK government soon after Kim Jong-un is no longer in power and before factions have a chance to develop. The factions might also be skeptical of the ROK government and assume that South Korea plans to dominate and quickly replace them. South Korea must assure the factions that it envisions an agreeable future for all North Korean elites, except the small minority who have been involved in serious human rights violations and related crimes.[10]

The *DPRK collapse: Negotiation* path would likely best be pursued by following the traditional route to unification involving a sequence of phases: (1) trust-building, (2) confederation (commonwealth), and (3) full unification. The trust-building phase would presumably involve South Korea and the United States providing considerable aid to North Korea, with particular focuses on humanitarian aid, infrastructure development, and capital investment. Ideally, the North would respond by developing national laws to protect ROK activities in the North and facilitating reasonable returns on investments while reducing the DPRK nuclear weapon and missile threats. The confederation phase would involve forming a shared, cooperative government, which might have relatively more ROK influence because of South Korea's economic might and population; however, the government would fundamentally need to compromise on a wide range of societal differences between the South and North to eventually facilitate full unification. Some of the issues that would require resolution would be property ownership, transitional justice, military force (including WMD) reductions, North Korean elite retention of politi-

---

[10] North Koreans who are not in the elite class will likely want all of the former elites removed from their positions and punished for their previous crimes. If South Korea takes this approach, it will have little chance of making progress toward unification because it will be threatening the very decisionmakers in the North who must decide to cooperate. Although, from a moral perspective, it would be ideal to empower those who had not been elites in North Korea, a more pragmatic approach will likely be required from a political and a capability perspective. For example, when the United States eliminated nearly all of the elites from the Iraqi government in 2003, the loss of capability, knowledge, and experience caused the Iraqi government to suffer many failures in the subsequent years.

cal positions and influence, and infrastructure maintenance and development. Although the ROK government has publicly disclosed very little about its plans to address these issues, my previous work provides recommendations for how each of these issues should be addressed.[11] South Korea is more likely to succeed in this path if it has appropriately prepared the policies and plans to take care of the North's elites. These policies need to be developed now and transmitted into North Korea so that the elites start to recognize that they might have an acceptable life after a post-Kim negotiated unification and begin to trust South Korea's plans.

As a confederation develops, many North Koreans will likely want to move to South Korea to take advantage of what they would perceive to be far better jobs and living conditions. But South Korea does not have the housing or sufficient jobs for the several million North Koreans who might attempt to move south. DPRK society could be clearly hurt by allowing professionals, such as doctors, to move to South Korea, and those at the lower end of the wage scale could form a massive flow of unskilled North Koreans, causing other disruptions in the South Korean economy.[12] The *DPRK collapse: Negotiation* path cannot form a unification that is simply a political resolution; it would also require a variety of policies and subsidized economic development in North Korea to avoid mass population flows.

Even if the DPRK government does not split into major factions, this path to unification will likely take many years. Trust does not develop overnight. Given the antagonism between the North and South, as well as North Korea's indoctrination of its people against South Korea, both sides can expect to be cautious and to try to avoid damaging surprises. As a rough estimate, trust-building might take three to five years, and confederation could easily last ten to 30 years as South Korea and the North Korean leaders develop relationships, build the North's economy and quality of life, and establish a degree of unity. Any degree of hostility that develops would jeopardize full uni-

---

[11] See, for example, Bennett, 2013; Bennett, 2017.

[12] ROK government authorities have told me that they are very concerned about the possibility of a massive flow of low-wage, unskilled workers to South Korea.

fication, and such hostility can be expected. In particular, even in this path, an insurgency of some magnitude can be expected that would take advantage of alienated military and security services personnel who would draw on munitions and other resources scattered in underground facilities across North Korea.[13]

Resolving these issues would be more complicated if the DPRK government split into multiple factions. Ultimately, South Korea would need to develop peaceful relations with the various factions and work gradually toward unification. South Korea and the United States would need to extend considerable humanitarian support to the various factions, as well as aid to help build their economies and repair or build new infrastructure. So that the infrastructures in each faction match, some degree of confederation would likely be required to provide needed power, water, roads, and other capabilities.

Given the challenges of confederation and the limited resources in North Korea in the best of times, factions developing in the country could still lead to some level of internal conflict and perhaps even civil war. As long as conditions in North Korea are stable and peaceful, China is less likely to intervene militarily.[14] But if China does intervene militarily, a ROK military intervention would also be likely, and this path would transition to the *DPRK collapse: Intervention* path. Such an outcome is more likely if a civil war develops in North Korea.

Conclusions on Paths to Unification Resulting from Regime Collapse

It is, of course, difficult to say how likely a DPRK regime collapse is in the coming years, although I believe it to be at least as likely as a major war on the Korean Peninsula. Today, the *DPRK collapse: Intervention* path seems more likely to occur than any of the three war-related paths or the *DPRK collapse: Negotiation* path, but with proper preparations,

---

[13] See Long, 2017.

[14] China will certainly intervene politically and economically if the DPRK government collapses.

the negotiation path could become a more likely path. The intervention path likely leads to a bad outcome because of the conflict and casualties that would likely result, especially from the use of North Korean WMD—although the casualties probably would not be as bad as the war-related paths to unification. The *DPRK collapse: Negotiation* path could lead to a good outcome, assuming that South Korea and the United States are effective in working with the successor government(s) after a DPRK regime collapse.

For the *DPRK collapse: Intervention* path, Chinese intervention and a failure to co-opt North Korean elites would have the most-serious impacts both for achieving an initial unification and for stabilizing whatever initial unification can be achieved. If China intervenes, South Korea would achieve only a partial initial unification. And even working with U.S. forces, South Korea might not be able to stabilize the part of North Korea that it controls—but it would need to do so to convince China that it can withdraw from the rest of the North without creating chaos and a power vacuum. Thus, some form of partial unification is the most likely medium- to longer-term outcome. If the ROK military is unable to stabilize the parts of the North that it occupies, many ROK citizens might begin arguing for withdrawing from the former North, given the cost of the ROK intervention (in both money and lives lost) and the low prospects of ever fully controlling all of Korea. If that happens, major parts of North Korea could be left in chaos and become ungovernable for many years, which would be a very bad outcome.

The *DPRK collapse: Negotiation* path is more hopeful. China seems much less likely to intervene if South Korea works peacefully with DPRK successor governments and especially if civil war in the North can be avoided (e.g., by starting the negotiation and trust-building process before factions form or soon thereafter). If China does intervene militarily, South Korea will almost certainly feel compelled to also intervene, causing the scenario to transition to the *DPRK collapse: Intervention* path. Without Chinese intervention, the pace of unification will be gradual, likely over several years, while trust and stability develop in North Korea and South Korea incrementally stabilizes the areas combined into the unified Korea.

CHAPTER FIVE

Peaceful Unification Paths

Both Koreas dream of unification, although the content of each dream appears to be quite different. Both seem to hope that a unified Korea might rise from a middle power to a major power and give Koreans a better place in the world. But the differences between the two Korean societies and governments would make a peaceful unification difficult to achieve. Indeed, a peaceful unification might proceed only as far as a confederation before the two sides recognize that one or the other is likely to be a loser in a unification: A win-win outcome for the two governments does not seem to exist.

In this chapter, I examine four peaceful paths to unification. But, first, I briefly review the literature on how German unification—a clearly peaceful path—could apply to Korean unification. This chapter describes the nature of the four paths, their likelihood, and their potential stability. These paths primarily differ in terms of which side controls the unification— North Korea, South Korea, or both (shared control). Each path assumes that, during the peaceful unification process, the Kim family regime remains in control of North Korea and a democratic government remains in control of South Korea. If the Kim family is overthrown en route to a unification, the scenarios would transition to the *DPRK collapse* paths discussed in the previous chapter.

German Unification: A Peaceful Example for Korea?

Many experts and politicians interested in Korean unification have sought lessons from the German unification experience, which has led to a very rich literature on the topic.[1] For many Koreans, the peaceful, negotiated German unification is viewed largely as a success that they would like to replicate. But the history of German unification suggests that, although it has been generally successful, it has also had its problems, especially for the East Germans. Thus, the literature applying the German unification case to Korea usually talks about mistakes that were made in Germany that should not be repeated in Korea.

But the more recent trend in this literature suggests that German unification was substantially different from a potential Korean case. For example, Rüediger Frank argues,

> A closer look, however, reveals that the differences between Germany and Korea far outweigh any similarities. This issue is not just an academic question; wrong assumptions can lead to wrong conclusions and to wrong policies. In the best case, such missteps would only waste money. In the worst case, however, they could lead to mismanagement of the unification process, with potentially disastrous consequences in the social, economic and security spheres.[2]

And Andrei Lankov writes,

> The calculations of the economists make us increasingly suspect that a German-style unification will disastrously break the South Korean economy, thus undermining the very basis of a unified country's prosperity.[3]

---

[1] See, for example, Kelly, 2011; Lankov, 2011; Marcus Noland, "Some Unpleasant Arithmetic Concerning Unification," Washington, D.C.: Peterson Institute for International Economics, Working Paper 96-13, 1996; Prantl and Hyun-Wook Kim, 2016; and Holger Wolf, "Korean Unification: Lessons from Germany," in Marcus Noland, ed., *Economic Integration of the Korean Peninsula*, Washington, D.C.: Peterson Institute for International Economics, January 1998.

[2] Rüediger Frank, "The Unification Cases of Germany and Korea: A Dangerous Comparison (Part 1 of 2)" 38 North, November 3, 2016.

[3] Andrei Lankov, "'Developmental Dictatorship' Could Be North Korea's Most Hopeful Future," Radio Free Asia, July 19, 2016.

In short, a peaceful unification of Korea, attempting to resemble the German reunification, could fail, which is one of the key issues for this report. As Lankov argues,

> I would like to start with a statement many people here will find disappointing and disturbing: there are virtually no reasons to expect that a negotiated and voluntary unification of the two Korean states is likely to happen in the foreseeable future. It has been noticed many times that the world history has no precedents of a negotiated peaceful and equal unification of two states. The only possible exception is Yemen, but in this country the supposedly "peaceful unification" was followed by a bitter civil war between the former North and former South—hardly an inspiring example. However, a closer look at Korea's situation indicates that in Korea as well the negotiated unification is not possible.[4]

In the remainder of this chapter, I examine four peaceful paths to Korean unification, all of which are worrisome for the reasons suggested here and otherwise.

Peace: DPRK Absorption Path

In Kim Jong-un's New Year's address in 2018, he spoke of Korean unification a dozen times.[5] In doing so, Kim was not speaking about unification by ROK absorption but rather about the kind of unification that North Korea has sought for decades. As Nicholas Eberstadt notes in his 1999 book,

> For the DPRK government, the reunification of Korea—on the DPRK's own terms—has been an overriding policy objective since its very inception. The urgent priority accorded to the goal of unconditional unification has been fused into the fundamental documents of both party and state. The preamble to the charter

---

[4] Andrei Lankov, "Unification and Great Powers," *Korea United*, December 2017.

[5] Kim Jong-un, 2018.

of the Korean Workers' Party (KWP) declares that "[t]he present task of the [KWP] is to ensure the complete victory of socialism in the Democratic People's Republic of Korea and the accomplishment of the revolutionary goals of national liberation and the people's democracy in the entire area of the country." And although the DPRK's seat of government has always been Pyongyang, the DPRK constitution from the outset stipulated that "the capital of the Democratic Republic of Korea shall be Seoul"—a claim whose realization would require the prior removal of the ROK government.[6]

The *War: DPRK conquest* path involves the DPRK military achieving a DPRK-controlled unification by invading South Korea. The *Peace: DPRK absorption* path reflects a North Korean subversion plan to control Korean unification via a political surrender of the ROK government to the DPRK government. For years, conservatives in South Korea have worried that the country's extreme progressives might attempt to achieve such an outcome. Still, most U.S. experts discount such a path, believing that the ROK populace would be too smart to allow it to happen.

Most ROK officials and experts argue that such an outcome is impossible today. But some outside experts fear that Kim Jong-un's subversive charm offensive that began in early 2018 appears intended to move South Korea in this direction—and to be making significant progress.[7] In particular, North Korea would first need to decouple the alliance between South Korea and the United States, and it is attempting to do so with the following strategies:

- North Korea is focusing attention on differences between the interests of the United States and those of South Korea. For example, DPRK nuclear weapons have posed a serious threat to

---

[6]   Nicholas Eberstadt, *The End of North Korea,* Washington, D.C.: American Enterprise Institute Press, 1999, Chapter One.

[7]   Of course, the current trade disputes between the United States and South Korea, the U.S.-ROK debate over how much South Korea should pay for the U.S. military presence there, and other alliance frictions only help North Korea in its efforts.

South Korea for many years, whereas the North's nuclear weapons pose a direct threat to the United States only once North Korea has an operational ICBM or similar nuclear weapon delivery means. The threat to South Korea was true throughout many of the years of U.S. President Barack Obama's administration, and yet, during that period, the U.S. strategy was referred to as "strategic patience"—trying to peacefully adapt to the growing North Korean threat by gradually strengthening U.S. and United Nations economic sanctions, deploying the Terminal High Altitude Area Defense missile defense system in South Korea, and perhaps taking some other actions.[8] Then, when the Trump administration came to office, it felt compelled to take further action because the United States could be targeted by North Korean ICBMs within a few years.

- Similarly, in 2018, ROK President Moon centered his strategy toward North Korea on peaceful coexistence, which is his personal preference for dealing with North Korea. At the same time, he knows that South Korea would likely be targeted by DPRK nuclear weapons if war developed, so preventing war by a peace initiative serves two purposes. But President Trump has threatened military attacks on the DPRK nuclear weapon program, which could well lead to war, to prevent North Korea from fielding nuclear weapons that could be used directly against the United States.

- For years, North Korea has pursued a peace treaty to end the Korean War. North Korea appears to believe that, once a peace treaty is completed, the U.S.-ROK alliance will be forced to dissolve (lacking a rationale), and U.S. military forces might well withdraw from South Korea.[9] Because the U.S. forces in South

---

[8]   In early 2017, *New York Times* reporters claimed that the United States had somehow hacked into the North Korean ballistic missiles and were causing some of the missile tests to fail. This, of course, is impossible to confirm at this time. See David E. Sanger, "A Eureka Moment for Two Times Reporters: North Korea's Missile Launches Were Failing Too Often," *New York Times*, March 6, 2017.

[9]   Trump's statement about his interest in eventually withdrawing U.S. forces from South Korea during the news conference of the June 12, 2018, Singapore Summit likely reassured

Korea provide significant support and infrastructure for the additional U.S. forces that would deploy to Korea in a conflict, withdrawing them would take away the follow-on forces' ability to promptly return, and they might not even try to return if war or some other form of DPRK dominance developed.

- In 2018, the terms North Korea used to describe its policy goal shifted from "peace treaty" to "peace agreement."[10] North Korea likely understands that a peace treaty would need to be approved by the U.S. Senate, where achieving the needed two-thirds vote would require Democratic and Republican consensus. And it would be difficult to achieve even a majority vote in the Senate unless the peace treaty included substantial provisions to create a true and lasting peace. North Korea wants a peace agreement that precedes establishing such conditions. Thus, a peace agreement that could be signed simply by the U.S. President appears to be the preferred North Korean approach.

- Kim Jong-un's superb information campaign has won major support for North Korea in South Korea. The United States has been far less effective in convincing the South Korean people about North Korea's failure to perform its promised actions.[11] As a

---

North Korea that this objective was possible. See Nancy Cook, Louis Nelson, and Nahal Toosi, "Trump Pledges to End Military Exercises as Part of North Korea Talks," *Politico*, June 12, 2018.

[10] For example, in 2015, it was reported that "North Korea on Saturday rejected the idea of resuming talks to end its nuclear program, saying previous such attempts ended in failure, and reiterated its demand that Washington come to the table to negotiate a peace treaty" (Jack Kim, "North Korea Rejects More Nuclear Talks, Demands Peace Treaty with U.S.," Reuters, October 17, 2015). The April 27 Panmunjom Declaration between ROK President Moon and Chairman Kim calls for "replacing the Armistice Agreement with a peace agreement and establishing a permanent and solid peace regime" (Moon Jae-in and Kim Jong-un, "Panmunjeom Declaration for Peace, Prosperity and Reunification of the Korean Peninsula," Panmunjeom, South Korea, April 27, 2018).

[11] For example, the April 27 Panmunjom Declaration calls for North and South Korea to fully implement "all existing inter-Korean declarations and agreements adopted thus far" (Moon and Kim, 2018). One of those declarations was the 1992 Denuclearization Declaration, in which the North and South committed not to "test, manufacture, produce, receive, possess, store, deploy or use nuclear weapons. South and North Korea shall use nuclear energy solely for peaceful purposes. South and North Korea shall not possess nuclear repro-

result, North Korea has made some progress in convincing the people of South Korea that the United States is an impediment to peace in Korea.

Some in South Korea believe that, even without U.S. military help, the ROK military could easily defeat the North Korean military in any conflict. But even if the conflict were limited to conventional warfare, the ROK Army, fighting without the U.S. Army, might struggle to defeat North Korea and would suffer substantial losses to DPRK artillery fire alone. In addition, South Korea appears to lack the logistical and supply capabilities required to defeat a DPRK invasion and then capture North Korea while destroying its military. And if North Korea uses its WMD, as it almost certainly would, the ROK military on its own would be challenged to defeat the DPRK military.

In short, a U.S. military withdrawal from South Korea would leave South Korea very vulnerable to North Korean coercion. That coercion could eventually lead to the ROK political leadership accepting peaceful coexistence under an arrangement dominated by North Korea. To prevent such an outcome, the ROK government might decide to develop its own nuclear weapons to offset the North Korean WMD threat in the event that the United States decides to withdraw its forces from South Korea. It would likely take several years for South Korea to build a nuclear force sufficient to offset North Korea's, and South Korea could be relatively vulnerable during this time.

Still, the *Peace: DPRK absorption* path has a very low likelihood of occurring. Even if North Korea could impose an initial, partial unification, the ROK populace would almost certainly reject this path and could well decide to fight against it. The path is likely possible only if the ROK-U.S. alliance is terminated and U.S. forces are withdrawn from the Korean Peninsula. It is highly likely that the ROK government will want to retain a solid alliance with the United States, which

---

cessing and uranium enrichment facilities" (Republic of Korea and Democratic People's Republic of Korea, "Joint Declaration of the Denuclearization of the Korean Peninsula," United Nations Department of Political Affairs, January 20, 1992).

is exactly what ROK President Moon has been saying.[12] Indeed, in late 2017, the ROK Minister of Unification, Cho Myoung-gyon, said as much about North Korea and its nuclear threat:

"Now that they are at the completion phase [of the nuclear program], they are coming up with new rhetoric that they haven't been emphasizing for a long time, like unifying the peninsula under a socialist regime," he said. Mr. Cho dismissed those aims as absurd. "I can say strongly and clearly that the unification that North Korea wants will never happen," he said.[13]

It is also important to note that the *Peace: DPRK absorption* path might lead to a unification that Kim Jong-un would not find favorable. If Kim suddenly controlled all of Korea, he would need vastly increased security service personnel to bring control to South Korea, an expansion of which would likely need to draw on KPA personnel, who are not as loyal and not as experienced as existing security service personnel. Kim would likely seek control by purging and probably killing many ROK elites, especially among the major businesses. Kim would want them eliminated because of their wealth and power, both of which he would want to absorb, and because of their international connections that he would probably want to cut to reduce the flow of information into Korea. Such cuts would disrupt the ROK economy, which is heavily oriented toward exports. Kim would want loyal followers controlling the ROK businesses, especially the major ones. But those loyal to him would not know how to run such businesses and, in many cases, would likely cause the businesses to fail, especially if international contacts and exports were restricted. Kim would need to cut ROK access to the internet and replace ROK computers with

---

[12] According to a *Korea Herald* report, "Cheong Wa Dae on Wednesday dismissed the idea of pulling U.S. troops out of South Korea following a peace treaty with North Korea, reiterating the need for a continued U.S. military presence on the Korean Peninsula" (Yeo Jun-suk, "USFK to Stay Even After Peace Treaty with NK: President Moon," *Korea Herald*, May 2, 2018).

[13] Jonathan Cheng and Andrew Jeong, "'The Unification That North Korea Wants Will Never Happen,' South Says," *Wall Street Journal*, November 16, 2017.

computers from North Korea that operate within the DPRK security restrictions. Because many ROK citizens would be purged while other survivors would lose their wealth, positions, international interactions, and computer connections, a major insurgency could develop in South Korea and likely spread into North Korea. Kim would clearly want to create a firewall between North Korea and South Korea because of the immense amount of information available and anger in South Korea, but this information and anger would still likely leak into North Korea, potentially destabilizing even the North Korean elites. There would be many risks that Kim Jong-un would be taking in trying to establish control over all of Korea, pushing him to consider an alternative approach, such as the *Peace: North Korea dominates* path.

Peace: ROK Absorption Path

German reunification was peaceful. West Germany effectively absorbed East Germany into a combined Germany, under control of what had been the West German leadership. The East German political leadership and Army were largely disenfranchised: After three years, only 2 percent of the East German military personnel who served before unification were still in the combined German military; 98 percent had been released to the civilian job market.[14]

As discussed earlier, German unification is often touted as a model for Korean unification. But German unification worked because the East German government was unable to control pressures from East German society for reform, and the government effectively lost the will to resist absorption into the West German government. It is unlikely that such pressures for reform among the common people would develop in North Korea.

Almost all experts believe that Kim Jong-un exercises strong control over North Korean society, such that major civil disobedience will not happen. But the Kim family appears to have worried about a dif-

---

[14] Dale R. Herspring, *Requiem for an Army: The Demise of the East German Military*, Lanham, Md.: Rowman and Littlefield Publishers, September 1998, pp. 182–183.

ferent example—the case of Nicolae Ceaușescu, the dictator of Romania, who was overthrown by his own military at the end of 1989. Kim Jong-un, in particular, has appeared to fear this case, having purged hundreds of senior DPRK personnel since he came to power in late 2011.[15] Kim has apparently taken more action against senior DPRK military officers in 2017 and 2018,[16] including replacing the three top military officers.[17]

In addition, Kim Jong-un brought his defense minister with him for the June 2018 Singapore Summit and, according to one source, "placed the country [North Korea] in heavy lockdown before leaving for the talks with the United States to prevent subversion, in other words, a political crisis that would threaten his rule."[18] Kim appears to be attempting to avoid a military coup that could lead to his having to negotiate a ROK-friendly unification.

If visible instability were to develop in North Korea, it is unlikely that Kim Jong-un would peacefully surrender to South Korea. Kim knows that South Korea would almost certainly arrest him for severe human rights violations and other crimes, and ROK courts would almost certainly find him guilty, leading to his execution. Even if South Korea does not take such action, international courts would almost certainly pursue Kim Jong-un. Although he could attempt to flee North Korea, it is unclear whether any country, even China, would provide him shelter from international legal actions against him.

---

[15] The think tank run by the ROK National Intelligence Service reported in December 2016 that "North Korea has purged a total of 340 people since leader Kim Jong-un took control of the communist country in 2011" ("N. Korea Purges 340 During 5-Year Rule of Kim Jong-un: Think Tank," Yonhap News Agency, December 29, 2016). The article appears to be referring to senior government officials removed from office, some of whom were killed.

[16] A former North Korean elite who has since left North Korea told me in July 2018 that, in 2017 and the first half of 2018, some 90 senior DPRK military officers were purged from their positions.

[17] "N. Korea Brings in Moderate as New Defense Minister," Yonhap News Agency, June 3, 2018.

[18] Ha Yoon Ah, "Regime Tightened Control over Key Personnel Prior to US-NK Summit," *DailyNK*, June 22, 2018.

Thus, the *Peace: ROK absorption* path seems extremely unlikely as well. Rebellion against Kim is not a high probability, and, if it occurs, he would almost certainly fight against ROK absorption—to the death, if need be. If the Kim family regime does not survive, the *Peace: ROK absorption* path would transition to one of the *DPRK collapse* paths, discussed in Chapter Four.

Peace: Cooperation Path

As noted in Chapter Four, South Korea has historically planned for a peaceful unification that would start with trust-building, transition to a confederation, and then move on to full unification.[19] Some level of trust-building is ongoing, and ROK President Moon has indicated a very strong interest in establishing a Korean confederation. It is possible that Kim Jong-un might also be interested in such an arrangement.

Conceptually, North and South Korea could enter into a confederation in which the vast majority of governmental functions (especially local government) remain separate, but some degree of coordination or integration is arranged between the North and South. This integration could then gradually increase over time, although it seems likely that it would take years and possibly decades before anything close to full unification would be possible along such a path. Of course, in the process, the Kim family could be overthrown once North Korean elites become more comfortable operating with the ROK government. In such circumstances, the scenario would transition to the *DPRK collapse: Negotiation* path discussed in Chapter Four.

The problem with a cooperation approach is the stark differences between South Korean and North Korean societies. South Korean society is based on democracy, free markets, and open internal and outside information. North Korea is an extreme dictatorship that has, only in desperation, allowed markets to develop illegally, and it is a

---

[19] In 2015, the ROK Ministry of Unification defined the three phases as (1) reconciliation and cooperation, (2) Korean commonwealth (confederation), and (3) completion of a unified nation. (ROK Ministry of Unification, 2015, p. 25).

closed society structured to prevent the North Korean people from being exposed to outside information. The Kim family accepted a compromise on having markets under a socialist economy in order to prevent starvation and economic failure in North Korea. Over the years, those markets and the activities associated with them have come to be a major force in the economy of North Korea, making any serious effort to eliminate them a likely cause of economic failure in North Korea. The market activities take a significant part of the control in North Korea away from the regime and give it to the entrepreneurs via the money they possess, allowing them to bribe DPRK officials so that they can break laws and rules. The regime has restricted outside information out of fears that the people would recognize what a poor leader Kim is, jeopardizing the survival of his regime. There does not appear to be a meaningful compromise between these two societal cultures: Any further opening of markets and information flow in the North tends to trample the DPRK regime's control and push the society in the direction of the ROK system. South Korea needs to begin information operations designed to move North Korean culture toward ROK culture, because waiting until a crisis or conflict will make mitigation much more difficult—maybe impossible.

These observations likely explain why President Moon has a strong preference for South-North interactions as envisioned in the 1991 Basic Agreement and then embodied in the April 27, 2018, Panmunjom Agreement. These interactions would encompass high-level diplomacy; economic exchanges; inter-Korean trade; exchanges in science, technology, education, arts, and sports; free intra-Korean travel and contacts; free correspondence, reunions, and visits; humanitarian issue resolution; railroad and road reconnections; opening of South-North sea and air routes; linking of postal and telecommunication services; joint economic and cultural undertakings abroad; and establishment of joint commissions (e.g., Economic Cooperation Commission).[20] Moon

---

[20] Parts of this wording come from the 1991 Basic Agreement (Republic of Korea and Democratic People's Republic of Korea, "Agreement on Reconciliation, Non-Aggression, and Exchanges and Cooperation Between South and North Korea," United Nations Department of Political Affairs, December 13, 1991).

likely wants these kinds of interactions because they would involve an opening of North Korea that would almost certainly draw North Korean elites and entrepreneurs toward the ROK societal framework, ultimately leading toward a unification based largely on ROK societal culture.

Although Kim Jong-un has agreed to eventually allow such interactions, he is starting very slowly, primarily with limited sports interactions. He is unlikely to accelerate much beyond the current efforts for fear of the outcome. He might have been willing to agree to these interactions to gain favor with the ROK people and government leaders, and this appearance of cooperation could lead to the *Peace: North Korea dominates* path (discussed next), which would significantly restrain North-South interactions. Regardless of his objective, Kim will likely develop concerns as these interactions proceed, and he will periodically reduce the extent of the interactions. His father, Kim Jong-il, used a similar approach in opening markets in North Korea, allowing some openings for a while and then reimposing restrictions for a while.[21] At some point, Kim could halt many of the interactions or even cancel the confederation.

In short, North Korea and South Korea may be prepared to peacefully enter a Korean confederation of some form. But a confederation is a serious risk for the DPRK regime because the momentum of change would more likely favor the ROK societal culture. Eventually, the Kim regime would most likely slow the confederation process and eventually stop it because the results of a balanced confederation favor South Korea and likely lead to a unification involving the end of the Kim regime and its influence on the Korean Peninsula.

Peace: North Korea Dominates Path

The *Peace: North Korea dominates* path is a variant of the *Peace: Cooperation* path. In it, the two Koreas enter a low-level confedera-

---

[21] See, for example, Barbara Demick, "North Korea Moves to Restrict Economy," *Los Angeles Times*, July 5, 2009.

tion, but almost all of the activities of the North and the South are still separate (a very loose confederation). Kim Jong-un would want to wait to form such a confederation until he has established a significant degree of enduring acceptance in South Korea.[22] North Korea would also likely prefer that this confederation develop after the U.S.-ROK alliance has been terminated and U.S. forces and other security guarantees are withdrawn from South Korea.[23] Moreover, when this confederation is established, North Korea would insist on having a very strong firewall between South Korea and North Korea—with severely limited interactions between the people of the two parts of the confederation (to control information flow). Simultaneously, North Korea would ramp up its indoctrination of its people, ever strengthening its criticism of the United States as an enemy to provide a scapegoat for regime failures and to discourage interest in democracy, free markets, and other strengths of the United States and its allies.

This path is different from the *Peace: Cooperation* path because it assumes that Kim Jong-un would have substantial leverage on the decisions made by the confederation. This leverage could be achieved if the North holds a majority on whatever body rules the confederation or if the ROK president allows Kim the leverage. This leverage would allow Kim to limit North-South interactions and sustain his control over North Korea. He could also insist that South Korea provide $50 bil-

---

[22] A poll in early May 2018 found that 77 percent of ROK citizens believed that Kim Jong-un is a leader worthy of trust. But that perspective developed very quickly and could also change quickly if Kim is recognized as having impaired peace on the peninsula or not been honest in the commitments he has made. The *Peace: North Korea dominates* path is much less likely if the ROK population comes to feel that Kim is not trustworthy. See Choe Sang-hun, "Kim Jong-un's Image Shift: From Nuclear Madman to Skillful Leader," *New York Times*, June 6, 2018.

[23] This appears to be one of the reasons that North Korea is so persistent in demanding a peace agreement with the United States. According to a *DailyNK* article,

> The reason that North Korea is focusing on the declaration of the end of the war while avoiding any real acts of denuclearization is so that it can succeed to dissolve the United Nations Command in South Korea which will thus lead to the withdrawal of the USKF [U.S. Forces Korea] while at the same time keeping their nuclear weapons intact. (Thae Yong Ho, "North Korea Seeks to Dissolve UN Command Through End-of-War Declaration," *DailyNK*, August 8, 2018)

lion to $100 billion each year to North Korea to subsidize its economic development. The Kim family would likely divert several billion dollars of this amount each year into its own coffers and use several billion dollars to provide for the desires of North Korean elites. The rest of the funding could then be used for economic development in the North, significantly rewarding North Korean capitalists, and especially those close to the regime.

From a DPRK perspective, this path may well be the best possible outcome. The Kim regime might hope that this kind of model could continue indefinitely, meeting the key needs of the Kim family and other North Korean elites while gradually improving the DPRK economy, making North Korea more independent and financially better off, and improving conditions for most of the rest of the population, thus making the regime more secure. Of course, purges, executions, and imprisonments, along with other harsh and barbaric control mechanisms, would need to be continued in the northern part of the confederation to sustain Kim family control. Kim Jong-un might also insist on some control mechanisms in the southern part of the confederation, especially if opposition developed to the annual subsidy given to the North (as would most likely be the case).

It is difficult to imagine this path leading to a true, lasting unification of Korea; if it did, it would most likely be a unification much like in the *Peace: DPRK absorption* path. But that path has considerable disadvantages for North Korea compared with this path. Still, the resulting confederation is unlikely to be stable. Even with a substantial firewall between North and South Korea, this kind of confederation would likely allow more information—and perhaps even a flood of information—from the outside to leak into North Korea. In addition, insurgency in South Korea could lead to military confrontation between the two sides. It is entirely possible that, if this configuration lasts ten years or so, the DPRK economy might be strengthened sufficiently to allow North Korea to once again be independent, leading the two sides to dissolve an uncomfortable confederation.

Conclusions on Peaceful Unification Paths

These four peaceful paths are very different from the paths that result from either war or DPRK regime collapse. Because, by definition, the peace context includes only paths in which both the Kim family regime and the democratic ROK government enter unification while in control of their countries, the use of force and its related challenges would presumably have little impact. The major issue determining the outcomes in these paths is whether Kim Jong-un would be willing to try any of the paths and whether he would stick with them. This is especially true of the *Peace: ROK absorption* path and the *Peace: Cooperation* path; it is really hard to imagine Kim Jong-un accepting either. It is also hard to imagine South Korea accepting the *Peace: DPRK absorption* path and then not rebelling against it. Of course, if an initial unification is somehow achieved in any of these paths, instability will likely mature over time and could doom the unification.

Only the *Peace: North Korea dominates* path appears relatively stable for as much as perhaps five to ten years, assuming that South Korea accepts it in the first place. Many experts would completely dismiss this path, asking why South Korea would ever accept a DPRK-dominated confederation. But this path seems to reflect the approach that Kim Jong-un is taking now with South Korea, trying to appear as a trustworthy leader who wants to eliminate the threat of war on the peninsula. Of course, this image neglects to recall that Kim Jong-un has caused the threat of war on the peninsula by his development of nuclear weapons and ballistic missiles, his tests of these weapons in contemptuous disregard of United Nations Security Council resolutions, his regular provocations, and his hostility and threats against the United States and South Korea so that within North Korea he can make the United States and South Korea the scapegoats for his regime's many failures. The probability of this path is difficult to determine. The probability is not high, but the path is possible.

CHAPTER SIX

Conclusions and Recommendations

Chapters Three through Five describe nine paths that could lead to Korean unification. None of these paths appears to have a high likelihood of success. The status quo of a separate North Korea and a separate South Korea appears more likely than a quick or even slow move to full or even partial unification. Although some of these paths resemble those discussed in the literature, there are also some significant differences.

In this chapter, I provide a summary comparison of the paths for each unification context that could develop in Korea and make recommendations for South Korea and the United States in each context. I then make overall recommendations for ROK and U.S. actions related to unification. A key issue across all paths is that neither South Korea nor North Korea is really ready for unification on any terms or through any process. They both appear to be approaching unification as largely a zero-sum game, thinking about unification as a case in which one side wins and the other loses. Kim Jong-un appears to be unwilling to consider a path in which he does not have a clear win in the unification process; instead, he appears to be pursuing the *Peace: North Korea dominates* path. And although ROK President Moon Jae-in appears to be more prepared for some form of unification compromise, he has yet to take the actions needed to make a unification that benefits both South and North Koreans. He needs to define policies for such a balanced approach.

Unification Resulting from a Major War

Chapter Three examined unification paths involving a major war. The three paths in this context include *War: DPRK conquest* of South Korea; *War: ROK-U.S. conquest—costly victory, contested peace*, which involves ROK and U.S. forces defeating North Korea and achieving a ROK-led unification; and *War: Improved ROK-U.S. conquest—reduced costs of victory, improved postconflict stability*, which involves South Korea and the United States setting conditions that would lead to a more favorable outcome than the second path.

Assessment

Most discussions of a major war on the Korean Peninsula do not adequately reflect the devastating impact that such a war could have. Even if a major Korean war involved the use of only conventional weapons, today's vastly enhanced conventional weapons could cause massive damage within the artillery range of the forward line of troops, as well as in areas affected by ballistic missiles, air strikes, cyberattacks, and special forces. But today and in the future, a major war in Korea would almost certainly involve a substantial use of WMD and, in particular, nuclear weapons. The economies and infrastructure of both North Korea and South Korea could be set back decades, with millions dead and seriously injured. Neither side wants such an outcome; therefore, both sides should seek to avoid unification through major war. This is especially true because such a unification would most likely be a partial unification, given the damage to the military forces and the likelihood of a significant Chinese intervention.

Still, there is some possibility of unification resulting from a major war. This likelihood takes into account the fact that both sides are very unlikely to start a major war, but war could still happen as a result of a miscalculation or accident. In particular, North Korea has a significant history of military provocations, and the United States has threatened military attacks against the DPRK nuclear weapon program. To deter these kinds of military attacks, both sides have threatened escalated responses that could spiral to a major war. For example, South Korea has adopted deterrence policies, such as its kill chain and Korean Massive Punishment and Retaliation strategy, that are highly escalatory and have a fairly significant potential of creating an escalation spiral to major war from a limited military exchange. On its side, North Korea has threatened to "destroy the Earth" if the North Korean regime is seriously threatened.

Recommendations for South Korea and the United States

South Korea and the United States need to enhance their deterrence of DPRK attacks. They have already postured a strong deterrent via the ROK concepts of proactive deterrence and Korean Massive Punishment and Retaliation, plus the U.S. Nuclear Posture Review's threat to destroy the North Korean regime if it ever uses a nuclear weapon. But South Korea and the United States have been reluctant to publicly repeat these threats in the midst of negotiating with North Korea. North Korea might perceive that these ROK and U.S. threats are a bluff and that the North's advancing nuclear capabilities will deter the execution of these threats. Instead, the ROK and U.S. deterrent threats should be made periodically to DPRK representatives in private meetings, accompanied by a logical explanation about why North Korea's nuclear capabilities would not prevent South Korea and the United States from acting on their threats to destroy the DPRK regime.

In addition, the United States should forgo its military threats against North Korea's nuclear weapon program. It is likely that the United States lacks the needed information to identify all of the significant facilities supporting that program. And even if the United States could identify these locations, the size of an effective U.S. attack would be quite large, likely ensuring a substantial retaliation by North Korea. Instead, if the North reneges on its commitments to dismantle its nuclear weapon program and surrender its nuclear weapons, the United States should threaten the regime with information operations. The former DPRK deputy ambassador to the United Kingdom, Thae Yong-ho, explained to the U.S. Congress in November 2017 that Kim Jong-un's regime survival could be significantly jeopardized by a flow of information into North Korea that makes it clear that Kim is not a god or an effective leader and that, in fact, he and his family have substantial failings.[1] Although such information operations could lead to some form of North Korean retaliation, they are far less likely to lead to a major war and could lead to regime failure.

Finally, South Korea and the United States need to more seriously contemplate the implications of a Chinese intervention into a major war with North Korea. Indeed, North Korea could decide to use nuclear weapons against China, which is certainly possible if China intervenes to achieve its own security and not to help North Korea against the United States. If North Korea does use such weapons against China, a Chinese nuclear retaliation against North Korea would be highly likely—and something the ROK and U.S. forces would want to avoid. Similarly, North Korea could execute nuclear attacks on South Korea and perhaps even the United States, but such attacks would also most likely lead to major nuclear retaliation against North Korea. South Korea and the United States need serious and frank coordination with the Chinese military to avoid nuclear collateral damage against each other, minimize the potential for military accidents, and provide a means for resolving accidents that most likely would occur sooner or later. In addition, the United States would want to seriously encourage South Korea not to consider trying to push Chinese forces out of North Korea but rather to seek a peaceful resolution (i.e., eventual withdrawal) of any Chinese military intervention.

Unification Resulting from a Regime Collapse

Chapter Four examined unification in the aftermath of a DPRK regime collapse. The two paths in this context include *DPRK collapse: Intervention*, which involves a DPRK regime failure that forces South Korea to intervene militarily in the North and unify as much of Korea as possible, and *DPRK collapse: Negotiation*, which involves a failure of the Kim family regime that leads to one or more factions taking control of the various parts of North Korea and establishing one or more successor governments.

---

[1]   Rachel Oswald, "North Korea Ripe for Info Campaign, Defector Tells House," *CQ News*, November 1, 2017.

Assessment

Many experts on North Korea believe that the North Korean regime is highly stable and that regime failure is very unlikely. But Kim Jong-un appears to have a different view of the regime's stability and his own security, exhibiting consistent paranoia. Examples include the killing of his uncle in 2013, the killing of his older brother by North Korean agents in 2017, his purges of dozens of senior elites each year,[2] and the way he locked down North Korea's elites while he traveled to meet with President Trump in June 2018.[3] More importantly, Kim is absolutely paranoid about outside information getting into North Korea, so he imposes severe prison sentences and even death for those who expose themselves to outside information.[4] And he has other substantial reasons to worry about regime survival, such as the dire economic conditions within North Korea (partly a result of sanctions) and the ire that his purges of North Korean elites have generated among their family and friends. This is not to argue that the DPRK regime will collapse tomorrow or anytime soon, but it could.

South Korea and the United States should thus be prepared to deal with the collapse of the regime, even if the probability of such a collapse is low. Much of the historical preparation for DPRK regime collapse has followed the *DPRK collapse: Intervention* path, assuming a ROK and likely U.S. military intervention in North Korea to impose

---

[2] As noted, a North Korean senior elite refugee told me in July 2018 that North Korea had purged about 90 senior military officers since early 2017, some of whom were executed. For the pattern before then, see "N. Korea Purges 340 During 5-Year Rule of Kim Jong-un: Think Tank," 2016.

[3] Ha, 2018.

[4] One of the most serious crimes that a North Korean can commit is to consume banned media. According to Freedom House, "listening to unauthorized foreign broadcasts and possessing dissident publications are considered 'crimes against the state'" in North Korea and "carry serious punishments, including hard labor, prison sentences, and the death penalty" (Freedom House, "North Korea," webpage, undated). On a single day in 2013, according to a report by a major South Korean newspaper, "the government executed 80 people in seven cities for violating such laws" (Jieun Baek, "The Opening of the North Korean Mind: Pyongyang Versus the Digital Underground," *Foreign Affairs*, January/February 2017). See also Kim Soo-hye and Lee Yong-soo, "N.Korean Teenagers Jailed for Listening to S.Korean Music," *Chosunilbo*, April 10, 2018.

a ROK-controlled unification. But this path has the potential of many of the same consequences as the *War: ROK-U.S. conquest—costly victory, contested peace* path, especially in an era of significant numbers of DPRK nuclear weapons and a regime that believes that it should try to "destroy the Earth" if its survival is jeopardized. Some of this risk could be reduced if South Korea establishes provisions for such a unification that would be friendly to North Korean elites. But because there appears to be little evidence that such preparations have been made, a stout North Korean resistance would likely be faced by the invading ROK and U.S. forces, even in a regime collapse context.

Because of the costs of warfare involving North Korea, the *DPRK collapse: Negotiation* path would be a preferable alternative to the *DPRK collapse: Intervention* path. The negotiation path would involve quickly working with the regime that replaces the Kim family to overcome the hostilities of the Kim regime and create a path toward a Korean confederation. Such an effort would require South Korea to formulate a concept for confederation and eventually unification that would be friendly to North Korean elites. This concept is more likely to be accepted by the North Koreans if it is formulated well before a regime collapse occurs. The downside of this path is that it would likely take several years to get to a serious confederation for Korea, and it would likely take many years thereafter to get to a full unification. But the upside would be that a gradual reeducation of the North Koreans would occur that could make a substantial reduction in the size of any insurgency against the confederation and the eventual full unification.

Recommendations for South Korea and the United States

South Korea should reconsider its approach to DPRK regime collapse. It should focus on what an effective post-collapse peaceful unification process requires and what aspects of such a process need to be put in place now. To facilitate this process, South Korea should begin developing the policies that would be applied to a confederation and eventual full unification, with particular emphasis on how North Korean elites would be treated under unification. These policies should then be widely broadcast into North Korea. More will be said about this process in the next section. Note that the policies for confederation and unification would be helpful not only in the *DPRK collapse: Negotiation* path but also in the *DPRK collapse: Intervention* path by encouraging North Korean elites and making it clear to everyone that the intervention path would not simply be a ROK absorption of North Korea.

Peaceful Unification

Chapter Five examined a peaceful unification of South Korea and the North Korean regime. The four paths in this context include *Peace: DPRK absorption*, in which North Korea absorbs the South; *Peace: ROK absorption*, in which South Korea absorbs the North; *Peace: Cooperation*, in which both sides form a balanced confederation; and *Peace: North Korea dominates*, in which the sides form a confederation, but North Korea dominates it.

Assessment

In this analysis, peaceful unification is interpreted as a decision to unify by the Kim family regime in North Korea and the democratic government of South Korea. (A peaceful unification in the aftermath of the collapse of North Korea is dealt with in the *DPRK collapse: Negotiation* path.) This framework significantly constrains the peaceful options available because Kim Jong-un does not appear to be prepared to accept a unification dominated by South Korea, an outcome that he knows would be a disaster for him personally. Instead, consistent with the efforts of his father and grandfather, Kim Jong-un seeks a peaceful unification controlled by North Korea. And Kim is quite anxious to achieve that outcome: In his 2018 New Year's speech, he emphasized the importance of achieving unification by discussing the subject 12 times.[5]

Most ROK and U.S. discussions of peaceful unification focus on a ROK-controlled unification, described in the *Peace: ROK absorption* path. It is impossible to believe that Kim Jong-un would accept such

---

[5]   Kim Jong-un, 2018.

an outcome. Instead, he might seek a DPRK-controlled unification. Some conservatives in South Korea worry that extreme progressives there would be prepared to surrender South Korea to Kim Jong-un, as described in the *Peace: DPRK absorption* path. Experts on South Korea think that this path is extremely unlikely, but several U.S. scholars are more concerned.[6] South Korea should take some action to reduce the likelihood of this possibility, even if it is not large.

In the literature, it is more common to talk of a shared unification in which Kim Jong-un and the president of South Korea jointly control the unification, as described in the *Peace: Cooperation* path. At first thought, this path might be considered much more possible. But the stark differences between North Korean and South Korean societies would make this path extremely difficult. Kim Jong-un would know that the much larger population of South Korea and the immensely larger ROK economy would tend to dominate North Korea in a unification with shared control, and it is difficult to believe that Kim would accept this outcome. More importantly, Kim would lose control of the subsequent flood of outside information into North Korea, which would persistently undermine his god-like image. He would thus want a confederation as opposed to a true unification, and he would seek to impose a substantial firewall against the flow of information and other things from South Korea. But for Kim, his difficulties in trying to sustain a firewall against South Korea despite complete control of North Korea would only be exacerbated by the confederation. Just as Kim Jong-il pulled back from allowing markets to develop in North Korea when he discovered his resulting loss of control, Kim Jong-un would likely pull back from a confederation when he discovered the degree to which it would undermine his control.

Still, there is a peaceful unification alternative, and Kim Jong-un appears to be pursuing it. In this alternative, the *Peace: North Korea dominates* path, Kim would pursue what appears to be a unification with shared control via confederation. But he would act to gain greater

---

[6]   Tara O has been an effective critic of the Moon government's actions. See, for example, Tara O, "Seoul Vulnerable: The Abandonment of the DMZ and the Destruction of South Korea's Military Capability," East Asia Research Center, September 27, 2018.

control of the confederation, imposing a substantial firewall between North and South Korea. He would apply minimal control over the ROK side of the confederation, seeking to avoid any rebellion against his control. His principal requirement of the ROK side of the confederation would be that it provide a substantial economic subsidy each year (perhaps $50 billion or $100 billion) to North Korea to ostensibly develop its economy, although the subsidy would also be used to meet the needs of the Kim family and North Korean elites, as well as the minimal needs (especially food) of all North Koreans. ROK opposition to even this limited DPRK control would likely build over time, and the DPRK firewall would also decay. Kim might eventually dissolve the confederation, after he substantially improves his economy, in order to sustain his control. Therefore, the *Peace: North Korea dominates* path is unlikely to lead to enduring unification.

Recommendations for South Korea and the United States

The difficulty of achieving any kind of real peaceful Korean unification is illustrated by these paths. They make it clear that the Kim family regime is the ultimate impediment to a real Korean unification. Unless other scholars are able to define a more fruitful peaceful unification path that could be accepted and followed by Kim Jong-un (which seems unlikely), those interested in Korean unification must begin thinking more about how to achieve relatively peaceful regime change in North Korea and to thereafter pursue something akin to the *DPRK collapse: Negotiation* path.

Simultaneously, South Korea and the United States need to help ROK citizens recognize the difficulty of a peaceful unification with the Kim family regime. South Koreans need to understand how disastrous it would be to accept a DPRK-controlled, so-called peaceful unification, which would not be very peaceful because of the purges that Kim Jong-un would undoubtedly want to impose on the leadership in South Korea. South Korea clearly also needs to more actively send information into North Korea to help the North Koreans understand democracy, market economics, and the good life available to all North Koreans if key changes are made.

Overall Recommendations

I conclude this analysis with three overall recommendations:

- South Korea must avoid using a major war to obtain unification, even in the case of a DPRK regime collapse; the cost would simply be too high. North Korea's nuclear and other WMD threats have changed the viability of war as a unification alternative. Still, a major war could occur, and South Korea needs to be prepared to prevent North Korea from defeating it and to achieve unification in case war does occur.
- South Korea needs to develop policies now that would provide most North Korean elites with a friendly outcome from unification.[7] Doing so is essential to achieving a solid initial unification and to making unification sustainable. If North Korean elites rebel against unification, unification could well fail. Once policies are developed, they need to be regularly communicated into North Korea using a fairly wide spectrum of communication approaches. For example, the ROK government should
  - change the charter of the Commission of the Five Provinces to make its appointed individuals the planned advisers to DPRK governors, mayors, and other officials rather than replacements for these officials. This change should be openly communicated to North Korea.
  - develop policies for military disarmament in the transition to unification. It is clear that a combined Korea should not have a military of more than 1.5 million military personnel (which is almost as large as the Chinese military). Initial military disarmament should employ the framework of reductions to equal force ceilings for both the North and South that North Korea proposed in the early 1990s (e.g., both North and

---

[7] Details on how to make unification friendly for North Korean elites can be found in Bennett, 2017.

South Korea would reduce their military forces to no more than 600,000 personnel).[8]

- South Korea and the United States should counter Kim Jong-un's image in North Korea as a god-like leader and his image in South Korea as a benevolent peacemaker. Kim is not a god-like leader, and he is not benevolent. He uses brutality to suppress perceived opposition and is responsible for incredibly heinous human rights violations against his people. He has made some progress with the DPRK economy, but much of the economic growth in North Korea has been due to North Korean capitalists. Kim's obsession with regime survival has driven him to waste scarce resources on a huge military, nuclear weapons, and other weapons that can actually endanger the North Korean people if ever used or if the nuclear weapon program is not dismantled as Kim has promised.
  - Kim is also not a peacemaker. His offer to abandon his nuclear weapons to achieve a summit with President Trump thus far appears to have been a classic bait and switch, with Kim unwilling to take any meaningful moves toward reducing, let alone abandoning, his nuclear weapons. Indeed, since offering to abandon his nuclear weapons, Kim has told his senior personnel that North Korea's nuclear weapons are key to regime survival,[9] suggesting that he has no intention of fully abandoning them. Kim claims he does not target the South Korean

---

[8] North Korea proposed military reductions to equal ceilings for both North and South Korea of 300,000 active-duty personnel in roughly 1992 (Yong-sup Han, "Peace and Arms Control on the Korean Peninsula," *International Journal of Korean Studies*, Vol. 5, No. 2, Fall/Winter 2001). The key question would be how to include security services and national police in such totals.

[9] According to a *DailyNK* report,

> On April 20th, one week before the Panmunjom Declaration was adopted, Kim Jong Un emphasized that North Korea's Nuclear weapons are "a powerful treasured sword for defending peace and the firm guarantee by which our descendants can enjoy the most dignified and happiest life in the world" at a meeting of the Central Committee of his ruling Workers' Party of Korea. In early July, the North Korean authorities gathered their core officials and held an internal lecture that emphasized that "nuclear weapons are a noble legacy left by former leaders Kim Il Sung and Kim Jong Il, and that if we do not have nuclear weapons, we die." (Thae Yong Ho, 2018)

people with nuclear weapons, yet many of his missiles are designed to deliver nuclear weapons there. And although Kim is seeking a peace agreement with the United States, he is refusing to create the conditions required for a permanent and solid peace as required by the April 27, 2018, Panmunjom Summit Declaration.[10]

— DPRK and ROK citizens need to better understand the realities of Kim's actions and objectives. South Korea and the United States should prepare significant information operations to convey this information.

— The threat of these U.S. actions should be used to induce desired North Korean behavior (such as freezing the building of new nuclear weapons), and implementation of these U.S. actions would be used as consequences imposed on North Korea in response to bad behavior—particularly North Korea's failure to perform its commitments, such as having no nuclear weapons or uranium enrichment capability.[11]

— Nevertheless, this psychological operation campaign needs to be properly organized and crafted. Experts should work diligently on this plan to help ensure that it does not backfire.

---

[10] The April 27 Declaration, item 3.3, says,

> South and North Korea agreed to actively pursue trilateral meetings involving the two Koreas and the United States, or quadrilateral meetings involving the two Koreas, the United States and China with a view to declaring an end to the War, turning the armistice into a peace treaty, and establishing a permanent and solid peace regime. (Moon Jae-in and Kim Jong-un, 2018)

A "permanent and solid peace" is not created by a peace agreement, especially given North Korea's history of violating its agreements. Permanent and solid peace is instead created by setting the conditions for peace, which North Korea refuses to do, suggesting that peace is not Kim's true intent.

[11] These commitments (and more) are included in the 1992 Denuclearization Declaration (Democratic People's Republic of Korea and Republic of Korea, 1992). In the April 27 Panmunjom Declaration, Moon Jae-in and Kim Jong-un committed their countries to fully implement all previous agreements and declarations (item 1.1), which includes the 1992 declaration.

If South Korea and the United States conclude that the preferred path for unification is the *DPRK collapse: Negotiation* path, they should commit to the survival of the DPRK regime and whoever leads it. And rather than simply waiting for a change in regime to happen, they could employ information operations on unification policies and on Kim Jong-un to help precipitate the downfall of the Kim family regime in favor of a successor regime. But such an approach would be risky for at least two reasons: Kim Jong-un could attempt to "destroy the Earth" when he senses his regime failing, and the nature of the successor regime cannot be predicted with certainty. Because Kim Jong-un severely restricts the flow of information into North Korea, it will be necessary to employ diverse and innovative means to spread information there.

The Challenge of North Korean Weapons of Mass Destruction

This appendix provides more details on the challenges inherent in North Korea possessing WMD, including nuclear, chemical, biological, and radiological weapons.

Nuclear Weapons

Table A.1 shows information on the six North Korean nuclear weapon tests that have been undertaken to date and how much physical damage these weapons could cause. The table shows the potential fatalities and serious injuries that could occur if a weapon of these yields is detonated as an airburst over the Yeouido subway station in Seoul.[1] Furthermore, for the blast and thermal radiation effects that can be lethal to at least some of the population, the table shows the radius of the area that would suffer serious effects. I chose a radius of 5 pounds per square

---

[1] Nuclear weapons detonated on the ground would vaporize a large amount of soil, rock, and other objects, lofting the vapor into the nuclear cloud. As the cloud cools, the vapor begins to solidify largely into small particles, carrying with them radiation also generated by the weapon. These solidified particles become fallout, which can also kill or injure many people and leave large areas contaminated and thus unusable for a period. To avoid fallout, nuclear weapons can be detonated at a high enough altitude to prevent the fireball from vaporizing many objects on the ground. These detonations are referred to as airbursts; if done at appropriate altitudes, an airburst can significantly enhance the blast effects of a nuclear weapon while avoiding the fallout contamination.

Table A.1
North Korean Nuclear Weapon Tests

Date of Test	Yield (Kt)[a]	Explosion at Yeouido		Effect Radius	
		Fatalities	Serious Injuries[b]	5-psi Blast	3rd-Degree Thermal Burns
October 2006	< 1	20,000	70,000	0.7 km	0.5 km
May 2009	4	50,000	180,000	1.1 km	1.0 km
February 2013	10	80,000	360,000	1.5 km	1.5 km
January 2016	6	60,000	240,000	1.3 km	1.2 km
September 2016	15	110,000	510,000	1.7 km	1.8 km
September 2017	250	820,000	2,500,000	4.4 km	6.6 km

SOURCES: For the yield data, see Bonnie Berkowitz and Aaron Steckelberg, "North Korea Tested Another Nuke. How Big Was It?" *Washington Post*, September 14, 2017. For the explosion and effect radius data, I calculated the numbers using Nukemap2; see Nuclear Secrecy, Nukemap2, online program, undated.

[a] A 1-kt weapon would cause damage comparable to a thousand tons of trinitrotoluene (or TNT).

[b] Serious injuries are in addition to the fatalities shown and would consist of radiation sickness, significant burns from thermal radiation, and significant physical damage from blasts effects. Not included would be minor injuries, such as minor cuts caused by shattered glass.

inch (psi) because, at that distance, "most buildings collapse," "injuries are universal, [and] fatalities are widespread."[2]

The fatality and serious injury numbers in Table A.1 are horrendous, especially for the most recent DPRK nuclear weapon test. But the effects would not end there. The psychological effects, in particular, would go well beyond the numbers shown in this table and, to at least some extent, would propagate through South Korea. These effects would be devastating, creating chaos throughout the country, especially because most people would expect further nuclear attacks, lead-

---

[2] R. Karl Zipf, Jr. and Kenneth L. Cashdollar, "Effects of Blast Pressure on Structures and the Human Body," Atlanta, Ga.: Centers for Disease Control and Prevention, undated. Nuclear blast effects are usually measured in terms of the psi of the shock wave. The 5-psi level is used in most analyses to indicate when buildings will collapse.

ing many people to evacuate ROK cities.[3] The South Korean economy could grind to a halt as workers failed to show up for work, and those who do go to work would likely find that the internet would not be working in at least some parts of the country, the suppliers of key goods had been destroyed, and the networks of a modern society had hemorrhaged—all from a single nuclear explosion in Seoul. Still, the specific outcomes are highly uncertain, depending on which people and facilities are affected and how badly, as well as how those effects propagate through South Korea.

If more than one nuclear weapon is used against South Korea, the numbers in Table A.1 could be multiplied by the number of detonations in cities, though showing somewhat decreasing marginal returns.

Potential Targeting of U.S. Military Bases in Korea

Chapter Two indicates that North Korea would likely target U.S. military bases in South Korea with nuclear weapons. One of the most senior DPRK military defectors, LTC Choi Ju-hwal, testified to the U.S. Congress in 1997 that

> Some Americans believe that even if North Korea possessed the ability to strike the United States, it would never dare to because of the devastating consequences. But I do not agree with this idea. If a war breaks out in the Korean Peninsula, the North's main target will be the U.S. forces based in the South and Japan. That is the reason why the North has been working furiously on its missile programs. Kim Jong-il believes that if North Korea creates more than 20,000 American casualties in the region, the U.S. will roll back and the North Korea will win the war.[4]

---

[3] I was a boy during the 1962 Cuban missile crisis in the United States. Just the serious threat of Soviet nuclear attack caused many people to empty store shelves and to evacuate U.S. cities. The fear of an attack was serious. In many cases, families stayed at home together.

[4] U.S. Senate, *North Korean Missile Proliferation: Hearing Before the Subcommittee on International Security, Proliferation, and Federal Services of the Committee on Governmental Affairs*, Washington, D.C., October 21, 1997, p. 5.

Ko Young-hwan, a former DPRK diplomat from the Ministry of Foreign Affairs, added,

> it is a well-known fact that North Korea will use short-range missiles and other missiles and rockets in order to have casualties of somewhere between 10,000 to 20,000, and even more casualties on the side of U.S. forces in order to have anti-war sentiments to rise inside the United States and cause the withdrawal of U.S. forces in the time of war.[5]

Thus, the North Korean regime might believe that early destruction of major U.S. military forces on the Korean Peninsula would defeat U.S. determination to continue fighting a major war and cause the United States to withdraw its forces from Korea because of the U.S. intolerance for significant casualties. Although most American deterrence experts perceive that any North Korean use of a nuclear weapon would likely lead to a disastrous nuclear exchange, the North Koreans might believe that nuclear weapons are usable in warfare. Indeed, the North Korean logic described here appears to parallel the Russian "escalate to deescalate" concept: Appropriate, early North Korean nuclear weapon use could lead to U.S. disengagement from a major Korean conflict and allow the North to then fight only South Korea, which would not have nuclear weapons for response.

Deterrents Against North Korean Nuclear Weapon Use

Since the late 1950s, the United States has offered a *nuclear umbrella* to South Korea. The nuclear umbrella promises South Korea that it does not need to have its own nuclear weapons. Instead, in circumstances requiring nuclear weapon use, the United States would provide the needed nuclear attacks. If North Korea detonated even one nuclear weapon on South Korea, South Korea could put significant pressure on the United States for a nuclear response that targets North Korea. And the 2018 U.S. Nuclear Posture Review states, "There is

---

[5]   U.S. Senate, 1997, p. 18. I heard a similar line of argument in a 2017 discussion with a DPRK military defector.

no scenario in which the Kim regime could employ nuclear weapons and survive."[6]

These threats do not explicitly say that the United States would use nuclear weapons in retaliating against any North Korean use of such weapons. But, given current limited U.S. capabilities to find and destroy the North Korean regime with conventional weapons, the United States would most likely have to use nuclear weapons. The United States would be less inclined to target DPRK cities, but many military targets are in or close to DPRK cities, especially Pyongyang. If the United States were to target Pyongyang consistent with the ROK Korean Massive Punishment and Retaliation strategy, a U.S. nuclear retaliation could destroy many of the DPRK government personnel who constitute the regime, leaving Kim Jong-un without the personnel needed for a functioning government even if he survived.

Chemical, Biological, and Radiological Weapons

How serious is the North Korean chemical weapon threat? In its killing of Kim Jong-un's older brother, Kim Jong-nam, with the VX nerve agent, North Korea illustrated that a very small quantity of this chemical weapon can be lethal, that North Korea possesses the weapon, and that it is willing to use it (even abroad). According to the ROK Ministry of National Defense, "North Korea began producing chemical weapons in the 1980s and currently holds a stockpile of an estimated 2,500 to 5,000 tons of chemical weapons."[7] If used effectively, this quantity of chemical weapons would cause substantial damage to the ROK military and likely to the ROK civilian population. Quantita-

---

[6] Office of the Secretary of Defense, 2018, p. 33.

[7] ROK Ministry of National Defense, 2016, p. 34. A former senior KPA officer with broad access to military information in the North told me that he heard from DPRK authorities that North Korea had 2,000 tons of chemical weapons. Another defector who held a very senior position in the DPRK military industrial complex told me in May 2017 that, in addition to what the North Koreans had made themselves, they had acquired at least hundreds of tons and perhaps as much as 2,000 tons of chemical weapons from a part of the former Soviet Union.

tively, one ton of sarin could kill tens of thousands of people in Seoul, and similar losses could be cause by 2 kg of anthrax if those exposed are left untreated.[8] Although some might question North Korea's ability to deliver a ton of chemical weapons to Seoul, a battery of the DPRK 240-mm rocket launchers (six launchers total) has the ability to deliver much more than a ton of chemical weapons to Seoul, and there appear to be dozens of such batteries within range of the ROK capital.

In 2004, when he was the U.S. Forces Korea Commander, GEN Leon LaPorte told the Senate Armed Forces Committee, "We also assess Pyongyang has an active biological weapons research program, with an inventory that may include anthrax, botulism, cholera, hemorrhagic fever, plague, smallpox, typhoid and yellow fever."[9] LaPorte did not describe how much of these agents might be available, and I am not familiar with other sources estimating the quantities.

The literature says little about DPRK radiological threats. These exist at least in terms of radioactive waste—likely in large quantities—from the DPRK nuclear weapon program and nuclear reactors.

---

[8]   These rough estimates are based on (1) the population density of Seoul in the Korean Statistical Information System and (2) the effect areas for chemical and biological agents (see Office of Technology Assessment, *Proliferation of Weapons of Mass Destruction: Assessing the Risks*, Washington, D.C.: U.S. Congress, OTA-ISC-559, August 1993, pp. 53–54; and Steve Fetter, "Ballistic Missiles and Weapons of Mass Destruction: What Is the Threat? What Should Be Done?" *International Security*, Vol. 16, No. 1, Summer 1991).

[9]   Leon J. LaPorte, statement before the Senate Armed Services Committee, Washington, D.C., April 1, 2004, p. 5.

References

Albert, Eleanor, "North Korea's Military Capabilities," Council on Foreign Relations, September 5, 2017. As of May 1, 2018:
https://www.cfr.org/backgrounder/north-koreas-military-capabilities

Auslin, Michael, "In Search of the Xi Doctrine," War on the Rocks, May 30, 2016. As of October 17, 2018:
https://warontherocks.com/2016/05/in-search-of-the-xi-doctrine

Baek, Jieun, "The Opening of the North Korean Mind: Pyongyang Versus the Digital Underground," *Foreign Affairs*, January/February 2017. As of October 17, 2018:
https://www.foreignaffairs.com/articles/north-korea/2016-11-28/opening-north-korean-mind

Bechtol, Bruce E., Jr., ed., *Confronting Security Challenges on the Korean Peninsula*, Quantico, Va.: Marine Corps University Press, 2011. As of October 17, 2018:
https://www.marines.mil/Portals/59/Publications/Confronting%20Security%20Challenges.%20On%20The%20Korean%20Peninsula.pdf

Bennett, Bruce W., *Preparing for the Possibility of a North Korean Collapse*, Santa Monica, Calif.: RAND Corporation, RAND, RR-331-SRF, 2013. As of October 17, 2018:
https://www.rand.org/pubs/research_reports/RR331.html

———, *Preparing North Korean Elites for Unification*, Santa Monica, Calif.: RAND Corporation, RR-1985-KOF, 2017. As of October 17, 2018:
https://www.rand.org/pubs/research_reports/RR1985.html

Bennett, Bruce W., and Jennifer Lind, "The Collapse of North Korea: Military Missions and Requirements," *International Security*, Vol. 36, No. 2, Fall 2011, pp. 84–119.

Berkowitz, Bonnie, and Aaron Steckelberg, "North Korea Tested Another Nuke. How Big Was It?" *Washington Post*, September 14, 2017. As of October 17, 2018:
https://www.washingtonpost.com/graphics/2017/world/north-korea-nuclear-yield

Bermudez, Jr., Joseph S., "Behind the Lines: North Korea's Ballistic Missile Units," *Jane's Intelligence Review*, June 14, 2011.

Cha, Victor, and David Kang, *Challenges for Korean Unification Planning: Justice, Markets, Health, Refugees, and Civil-Military Transitions*, Washington, D.C.: USC Korean Studies Institute and Center for Strategic and International Studies, December 2011. As of October 17, 2018: https://csis-prod.s3.amazonaws.com/s3fs-public/legacy_files/files/publication/120110_Cha_ChallengesKorea_WEB.pdf

Cheng, Jonathan, and Andrew Jeong, "'The Unification That North Korea Wants Will Never Happen,' South Says," *Wall Street Journal*, November 16, 2017.

Choe Sang-hun, "Kim Jong-un's Image Shift: From Nuclear Madman to Skillful Leader," *New York Times*, June 6, 2018. As of October 17, 2018: https://www.nytimes.com/2018/06/06/world/asia/kim-korea-image.html

Cook, Nancy, Louis Nelson, and Nahal Toosi, "Trump Pledges to End Military Exercises as Part of North Korea Talks," *Politico*, June 12, 2018.

Demick, Barbara, "North Korea Moves to Restrict Economy," *Los Angeles Times*, July 5, 2009. As of October 17, 2018: http://articles.latimes.com/2009/jul/05/world/fg-north-korea-economy

Denyer, Simon, and Amanda Erickson, "Beijing Warns Pyongyang: You're on Your Own If You Go After the United States," *Washington Post*, August 11, 2017. As of October 17, 2018: https://www.washingtonpost.com/world/china-warns-north-korea-youre-on-your-own-if-you-go-after-the-us/2017/08/11/a01a4396-7e68-11e7-9026-4a0a64977c92_story.html

"Dresden Initiative for Peaceful Unification on the Korean Peninsula," KBS World Radio, March 28, 2014. As of October 17, 2018: http://world.kbs.co.kr/special/kdivision/english/tasks/dresden.htm

Eberstadt, Nicholas, *The End of North Korea*, Washington, D.C.: American Enterprise Institute Press, 1999.

Fetter, Steve, "Ballistic Missiles and Weapons of Mass Destruction: What Is the Threat? What Should Be Done?" *International Security*, Vol. 16, No. 1, Summer 1991, pp. 5–42.

Frank, Rüediger, "The Unification Cases of Germany and Korea: A Dangerous Comparison (Part 1 of 2)" 38 North, November 3, 2016. As of October 17, 2018: https://www.38north.org/2016/11/rfrank110316/

Freedom House, "North Korea," webpage, undated. As of October 17, 2018: https://freedomhouse.org/report/freedom-press/2012/north-korea

Gale, Alastair, and Kwanwoo Jun, "South Korea's Governors of Northern Provinces Don't—and Never Will—Govern," *Wall Street Journal*, March 17, 2014. As of October 17, 2018: http://online.wsj.com/article/SB10001424052702304419104579321810508073546.html

Garamone, Jim, "US-South Korean Alliance Ready to Defend Against North Korean Threat, Top Generals Say," DoD News, Defense Media Activity, August 15, 2017. As of October 17, 2018:
https://www.army.mil/article/192359/us_south_korean_alliance_ready_to_defend_against_north_korean_threat_top_generals_say

Glaser, Bonnie, Scott Snyder, and John S. Park, *Keeping an Eye on an Unruly Neighbor: Chinese Views of Economic Reform and Stability in North Korea*, Washington, D.C.: Center for Strategic and International Studies and U.S. Institute of Peace, January 3, 2008.

Ha Yoon Ah, "Regime Tightened Control Over Key Personnel Prior to US-NK Summit," *DailyNK*, June 22, 2018. As of October 17, 2018:
http://www.dailynk.com/english/regime-tightened-control-over-key-personnel-prior-to-us-nk-summit

Han, Yong-sup, "Peace and Arms Control on the Korean Peninsula," *International Journal of Korean Studies*, Vol. 5, No. 2, Fall/Winter 2001, pp. 53–70. As of October 17, 2018:
http://icks.org/data/ijks/1482456353_add_file_4.pdf

Helvey, David F., *Korean Unification and the Future of the U.S.-ROK Alliance*, Washington, D.C.: National Defense University, Strategic Forum 291, February 2016. As of October 17, 2018:
http://www.dtic.mil/dtic/tr/fulltext/u2/1004294.pdf

Herspring, Dale R., *Requiem for an Army: The Demise of the East German Military*, Lanham, Md.: Rowman and Littlefield Publishers, September 1998.

"Is Peaceful Korean Unification Possible?" editorial, *New York Times*, December 11, 2014. As of October 17, 2018:
https://www.nytimes.com/2014/12/12/opinion/is-peaceful-korean-unification-possible.html

Jun Hyun-suk, "Troops to Be Slashed to 500,000 by 2022," *Chosunilbo*, January 22, 2018. As of October 17, 2018:
http://english.chosun.com/site/data/html_dir/2018/01/22/2018012201244.html

Kelly, Robert, "The German-Korean Unification Parallel," *Korean Journal of Defense Analysis*, Vol. 23, No. 4, December 2011, pp. 457–472.

Kim Dae-jung, "Inaugural Address," Ch'ongwadae, South Korea, February 25, 1998.

Kim Eun-jung, "S. Korea, U.S. Agree to Set N. Korean Nuclear Deterrence Policy by 2014," Yonhap News Agency, October 25, 2012. As of October 17, 2018:
http://english.yonhapnews.co.kr/national/2012/10/24/58/0301000000AEN20121024006651315F.HTML

Kim Hyun Sik, "The Secret History of Kim Jong Il," *Foreign Policy*, September/October 2008. As of October 17, 2018: http://foreignpolicy.com/2009/10/06/the-secret-history-of-kim-jong-il

Kim, Jack, "North Korea Rejects More Nuclear Talks, Demands Peace Treaty with U.S.," Reuters, October 17, 2015. As of October 17, 2018: https://www.reuters.com/article/ us-northkorea-usa-nuclear-idUSKCN0SB0QN20151017

Kim Jong-un, "New Year's Address," NK Leadership Watch, January 1, 2018. As of October 17, 2018: http://www.nkleadershipwatch.org/2018/01/01/new-years-address

Kim Myong-sik, "'Unification Bonanza' Is Misleading Slogan," *Korea Herald*, February 5, 2014. As of October 17, 2018: http://www.koreaherald.com/view.php?ud=20140205000555

Kim Soo-hye and Lee Yong-soo, "N.Korean Teenagers Jailed for Listening to S.Korean Music," *Chosunilbo*, April 10, 2018. As of October 17, 2018: http://english.chosun.com/site/data/html_dir/2018/04/10/2018041000751.html

"Korea-China Summit," editorial, *Korea Herald*, March 25, 2014. As of October 17, 2018: http://www.koreaherald.com/view.php?ud=20140325000519

Korean Statistical Information Service, "Projected Population by Age (Korea)," population database, undated. As of October 17, 2018: http://kosis.kr/eng

Lankov, Andrei, "A Legal Minefield for Korean Reunification," *Asia Times*, July 30, 2011. As of October 17, 2018: http://www.atimes.com/atimes/Korea/MG30Dg01.html

———, "'Developmental Dictatorship' Could Be North Korea's Most Hopeful Future," Radio Free Asia, July 19, 2016. As of October 17, 2018: https://www.rfa.org/english/commentaries/parallel-thoughts/ korea-development-07192016155616.html

———, "Unification and Great Powers," *Korea United*, December 2017. As of October 17, 2018: http://www.koreaunited.kr/wp-content/uploads/2017/12/Andrei-Lankov.pdf

LaPorte, Leon J., statement before the Senate Armed Services Committee, Washington, D.C., April 1, 2004. As of October 17, 2018: ogc.osd.mil/olc/docs/test04-04-01Laporte.doc

Lee, Byeonggu, "The Role of the Republic of Korea-U.S. Alliance in Peaceful Unification of Korea," *International Journal of Korean Studies*, Vol. 19, No. 2, Fall 2015, pp. 47–68.

Lee, Sung-yoon, "North Korea's Revolutionary Unification Policy," *International Journal of Korean Studies*, Vol. 18, No. 2, Fall 2014, pp. 121–137.

Long, Austin, *Insurgency in the DPRK? Post-Regime Insurgency in Comparative Perspective*, Baltimore, Md.: U.S.-Korea Institute Paul H. Nitze School of Advanced International Studies, Johns Hopkins University, March 2017.

Martina, Michael, "China Won't Allow Chaos or War on Korean Peninsula: Xi," Reuters, April 28, 2016.

Mastro, Oriana Skylar, "Will China Invade North Korea and Take Its Nuclear Facilities?" *Newsweek*, September 14, 2017. As of October 17, 2018:
http://www.newsweek.com/
will-china-invade-north-korea-and-take-its-nuclear-facilities-665008

Maxwell, David S., "A Strategy for the Korean Peninsula: Beyond the Nuclear Crisis," *Military Review*, September/October 2004.

———, "Should the United States Support Korean Unification and If So, How?" *International Journal of Korean Studies*, Vol. 18, No. 1, Spring 2014, pp. 139–156.

———, "Unification Options and Scenarios: Assisting a Resistance," *International Journal of Korean Unification Studies*, Vol. 24, No. 2, 2015, pp. 127–152.

Mclaughlin, Kelly, "China 'Deploys 150,000 Troops to Deal with Possible North Korean Refugees over Fears Trump May Strike Kim Jong-un Following Missile Attack on Syria,'" *Daily Mail*, April 10, 2017. As of October 17, 2018:
http://www.dailymail.co.uk/news/article-4399076/
China-deploys-150-000-troops-North-Korea-border.html

Menon, Prakash, and P. R. Shankar, "Could North Korea Destroy Seoul with Its Artillery Guns?" *National Interest*, May 25, 2018. As of October 17, 2018:
https://nationalinterest.org/blog/the-buzz/
could-north-korea-destroy-seoul-its-artillery-guns-25963

Moon Jae-in and Kim Jong-un, "Panmunjeom Declaration for Peace, Prosperity and Reunification of the Korean Peninsula," Panmunjeom, South Korea, April 27, 2018. As of October 17, 2018:
http://english1.president.go.kr/BriefingSpeeches/Speeches/32

"N. Korea Brings in Moderate as New Defense Minister," Yonhap News Agency, June 3, 2018. As of October 17, 2018:
http://english.yonhapnews.co.kr/northkorea/2018/06/03/
0401000000AEN20180603000951315.html

"N. Korea Purges 340 During 5-Year Rule of Kim Jong-un: Think Tank," Yonhap News Agency, December 29, 2016. As of October 17, 2018:
http://english.yonhapnews.co.kr/northkorea/2016/12/29/68/
0401000000AEN20161229003600315F.html

Noland, Marcus, "Some Unpleasant Arithmetic Concerning Unification," Washington, D.C.: Peterson Institute for International Economics, Working Paper 96-13, 1996.

Nuclear Secrecy, Nukemap2, online program, undated. As of October 17, 2018:
http://nuclearsecrecy.com/nukemap

O, Tara, "Seoul Vulnerable: The Abandonment of the DMZ and the Destruction of South Korea's Military Capability," East Asia Research Center, September 27, 2018. As of October 17, 2018:
https://eastasiaresearch.org/2018/08/18/seoul-vulnerable-the-abandonment-of-the-dmz-and-the-destruction-of-south-koreas-military-capability

Office of the Secretary of Defense, *2018 Nuclear Posture Review*, Washington, D.C.: U.S. Department of Defense, February 2018. As of October 17, 2018:
https://media.defense.gov/2018/Feb/02/2001872886/-1/-1/1/2018-Nuclear-Posture-Review-Final-Report.pdf

Office of Technology Assessment, *Proliferation of Weapons of Mass Destruction: Assessing the Risks*, Washington, D.C.: U.S. Congress, OTA-ISC-559, August 1993.

Oswald, Rachel, "North Korea Ripe for Info Campaign, Defector Tells House," *CQ News*, November 1, 2017.

Oxford University Press, English: Oxford Living Dictionaries, web tool, undated. As of October 17, 2018:
https://en.oxforddictionaries.com/definition/capitalist

Panda, Ankit, "US Intelligence: North Korea May Already Be Annually Accruing Enough Fissile Material for 12 Nuclear Weapons," *The Diplomat*, August 9, 2017. As of October 17, 2018:
https://thediplomat.com/2017/08/us-intelligence-north-korea-may-already-be-annually-accruing-enough-fissile-material-for-12-nuclear-weapons

Park Geun-hye, "A New Kind of Korea," *Foreign Affairs*, September/October 2011. As of October 17, 2018:
https://www.foreignaffairs.com/articles/northeast-asia/2011-09-01/new-kind-korea

Park Won Gon, "Strategic Implications of the USFK Relocation to Pyeongtaek," *ROKAngle*, No. 164, Korea Institute for Defense Analyses, October 20, 2017. As of October 17, 2018:
www.kida.re.kr/cmm/viewBoardImageFile.do?idx=23487

Peck, Michael, "North Korea Plans to Defeat the U.S. Army in a War. Here's How," *National Interest*, January 12, 2018. As of October 17, 2018:
https://nationalinterest.org/blog/the-buzz/north-korea-plans-defeat-the-us-army-war-heres-how-24029

Pollack, Jonathan D., and Chung Min Lee, *Preparing for Korean Unification: Scenarios and Implications*, Santa Monica, Calif.: RAND Corporation, MR-1040-A, 1999. As of October 17, 2018:
https://www.rand.org/pubs/monograph_reports/MR1040.html

Prantl, Jochen, and Hyun-wook Kim, "Germany's Lessons for Korea: The Strategic Diplomacy of Unification," *Global Asia*, December 27, 2016. As of October 17, 2018:
https://www.globalasia.org/bbs/board.php?bo_table=articles&wr_id=9161

Republic of Korea, "Constitution of the Republic of Korea," National Assembly of the Republic of Korea, October 29, 1987. As of October 17, 2018:
http://korea.assembly.go.kr/res/low_01_read.jsp?boardid=1000000035

Republic of Korea and Democratic People's Republic of Korea, "Agreement on Reconciliation, Non-Aggression, and Exchanges and Cooperation Between South and North Korea," United Nations Department of Political Affairs, December 13, 1991. As of October 17, 2018:
https://peacemaker.un.org/korea-reconciliation-nonaggression91

———, "Joint Declaration of the Denuclearization of the Korean Peninsula," United Nations Department of Political Affairs, January 20, 1992. As of October 17, 2018:
https://peacemaker.un.org/korea-denuclearization92

Republic of Korea Ministry of National Defense, *2000 Defense White Paper*, Seoul, 2000.

———, *2016 Defense White Paper*, Seoul, December 31, 2016. As of October 17, 2018:
http://www.mnd.go.kr/user/mndEN/upload/pblictn/
PBLICTNEBOOK_201705180357180050.pdf

Republic of Korea Ministry of Unification, *Sunshine Policy for Peace and Cooperation*, Seoul, May 2002.

———, *Initiative for Korean Unification*, Seoul, October 2015.

———, *2016 White Paper on Korean Unification*, Seoul, May 2016.

———, "Moon Jae-In's Policy on the Korean Peninsula: A Peninsula of Peace and Prosperity," Seoul, 2017.

Revere, Evans J. R., "Korean Reunification and U.S. Interests: Preparing for One Korea," Washington, D.C.: Brookings Institution, January 20, 2015. As of October 17, 2018:
https://www.brookings.edu/on-the-record/
korean-reunification-and-u-s-interests-preparing-for-one-korea

ROK Ministry of National Defense—*See* Republic of Korea Ministry of National Defense.

ROK Ministry of Unification—*See* Republic of Korea Ministry of Unification.

"S. Korea Unveils Plan to Raze Pyongyang in Case of Signs of Nuclear Attack," Yonhap News Agency, September 11, 2016. As of October 17, 2018: http://english.yonhapnews.co.kr/search1/ 2603000000.html?cid=AEN20160911000500315

Sanger, David E., "A Eureka Moment for Two Times Reporters: North Korea's Missile Launches Were Failing Too Often," *New York Times*, March 6, 2017. As of October 17, 2018: https://www.nytimes.com/2017/03/06/insider/a-eureka-moment-for-two-times-reporters-north-koreas-missile-launches-were-failing-too-often.html

Ser Myo-ja, "Park Tells Military to Strike Back If Attacked," *Korea JoongAng Daily*, April 2, 2013. As of October 17, 2018: http://koreajoongangdaily.joins.com/news/article/article.aspx?aid=2969490

Song Sang-ho, "The Cut Will Come in Sync with Plans to Pare Down the Number of Active-Duty Troops to 500,000 by 2022 from the Current 618,000," Yonhap News Agency, July 27, 2018. As of October 17, 2018: http://english.yonhapnews.co.kr/search1/ 2603000000.html?cid=AEN20180727003900315

Terry, Sue Mi, *Unified Korea and the Future of the U.S.-South Korea Alliance*, New York: Council on Foreign Relations, December 2015. As of October 17, 2018: https://cfrd8-files.cfr.org/sites/default/files/pdf/2015/12/ Discussion_Paper_Korea_Unification_Terry.pdf

Thae Yong Ho, "North Korea Seeks to Dissolve UN Command Through End-of-War Declaration," *DailyNK*, August 8, 2018. As of October 17, 2018: https://www.dailynk.com/english/ north-korea-seeks-to-dissolve-un-command-through-end-of-war-declaration

"U.S. General Says Forces Ready to Counter N.Korean Attack," *Chosunilbo*, July 15, 2009.

U.S. Senate, *North Korean Missile Proliferation: Hearing Before the Subcommittee on International Security, Proliferation, and Federal Services of the Committee on Governmental Affairs*, Washington, D.C., October 21, 1997. As of October 17, 2018: https://www.gpo.gov/fdsys/pkg/CHRG-105shrg44649/pdf/ CHRG-105shrg44649.pdf

————, *Fiscal Year 2006 Defense Department Budget: Hearing of the Senate Armed Services Committee*, Washington, D.C., U.S. Government Printing Office, March 8, 2005.

Vincenzo, Fredrick, *An Information Based Strategy to Reduce North Korea's Increasing Threat: Recommendations for ROK & U.S. Policy Makers*, Washington, D.C.: Center for a New American Security, Georgetown University, National Defense University, and U.S.-Korea Institute at SAIS, October 2016. As of October 17, 2018:
http://georgetownsecuritystudiesreview.org/wp-content/uploads/2016/10/An-Information-Based-Strategy-to-Reduce-North-Koreas-Increasing-Threat.pdf

"What North and South Korea Would Gain If They Were Reunified," *The Economist*, May 5, 2016. As of October 17, 2018:
https://www.economist.com/graphic-detail/2016/05/05/what-north-and-south-korea-would-gain-if-they-were-reunified

Wolf, Holger, "Korean Unification: Lessons from Germany," in Marcus Noland, ed., *Economic Integration of the Korean Peninsula*, Washington, D.C.: Peterson Institute for International Economics, January 1998.

Yeo Jun-suk, "USFK to Stay Even After Peace Treaty with NK: President Moon," *Korea Herald*, May 2, 2018. As of October 17, 2018:
http://www.koreaherald.com/view.php?ud=20180502000177

Zipf, R. Karl, Jr., and Kenneth L. Cashdollar, "Effects of Blast Pressure on Structures and the Human Body," Atlanta, Ga.: Centers for Disease Control and Prevention, undated. As of October 17, 2018:
https://www.cdc.gov/niosh/docket/archive/pdfs/NIOSH-125/125-ExplosionsandRefugeChambers.pdf